Geology and Health:

Closing the Gap

Geology and Health:

Closing the Gap

EDITED BY:

H. Catherine W. Skinner
Department of Geology and Geophysics
Yale University
Department of Orthopaedics & Rehabilitation
Yale School of Medicine

AND

Antony R. Berger
Victoria, British Columbia
Canada

With assistance from the
International Union of Geological Sciences and the
International Geological Correlation Programme

New York   Oxford
OXFORD UNIVERSITY PRESS
2003

OXFORD
UNIVERSITY PRESS

Oxford   New York
Auckland   Bangkok   Buenos Aires   Cape Town   Chennai
Dar es Salaam   Delhi   Hong Kong   Istanbul   Karachi   Kolkata
Kuala Lumpur   Madrid   Melbourne   Mexico City   Mumbai   Nairobi
São Paulo   Shanghai   Taipei   Tokyo   Toronto

Published by Oxford University Press, Inc.
198 Madison Avenue, New York, New York 10016

www.oup.com

Oxford is a registered trademark of Oxford University Press

Library of Congress Cataloging-in-Publication Data
Skinner, H. Catherine W.
Geology and health : closing the gap / H. Catherine W. Skinner and Antony R. Berger.
 p. cm.
Includes bibliographical references and index.
ISBN 0-19-516204-8
1. Environmentally induced diseases. 2. Environmental health. 3. Medical geography.
4. Geology—Health aspects. I. Berger, Antony R. II. Title
RA566 .S556 2003
614.4'2—dc21   2002032955

9 8 7 6 5 4 3 2 1

Printed in the United States of America
on acid-free paper

Acknowledgments

Funds to bring researchers from many countries to Uppsala were provided by the International Union of Geological Sciences Working Group on Medical Geology, the International Geological Correlation Programme (via Project #454), the UNESCO-ICSU-IUGS funded project "Paracelsus Revisited," The Center for Metal Biology in Uppsala, The Royal Swedish Academy of Sciences, The Swedish Natural Science Research Council, and the Swedish Institute. For a summary report on the meeting see <http://www.swipnet.se/medicalgeology/workshop_in_uppsala.htm>.

While taking full responsibility for the choice of the manuscripts included in this volume, we are greatly indebted to the following reviewers, who offered technical and general comments on the presentations and the whole publication project:

In the U.S.:
J.A. Albright, Shreveport, LA, S. Bohlen, Washington, DC; T. S. Bowers, Cambridge, MA, O. Bricker, Reston, VA, D. Carter, New Haven, CT, R. Coleman, Palo Alto, CA, W. G. Ernst, Stanford, CA, R.B. Gordon, New Haven, CT, B. Haley, Lexington, KY, L.J. Hickey, New Haven, CT, M. Hindelang, Houghton, MI, B.F. Jones, Reston, VA, R.J. Kamilli, Tucson, AZ, K.D. Kupperman, New York, NY, P. Leahy, Reston, VA, C. Palmer, Reston, VA, S. Porter, Seattle, WA; M. Ross, Washington, DC, J Samburg, New Haven, CT, I. Shapiro, Philadelphia, PA, B.J. Skinner, New Haven, CT, Sourkov, New Haven, CT, S.T. Thompson, Norwalk, CT, J.D. Webster, New York, NY.

In Canada:
C. Dunn, Sydney, BC, D.C. Elliott, Calgary, AB, A.M. Glen, Victoria, BC R.E. Lett, Victoria, BC, N.W.D. Massey, Victoria, BC, S. Paradis, Sidney, BC, B.A. Roberts, Corner Brook, NF, B. Sherriff, Winnipeg, MN.

In the U.K.:
J.D.Appleton, Nottingham; N. Breward, Nottingham; F. Fordyce, Edinburgh; J.S.G., McCullock, Abington; B. Rawlins, Nottingham; P. Smedley, Nottingham; A. Stewart, Birkenhead; P.G. Whitehead, Reading.

The editors wish to express their sincere appreciation to all the authors for their excellent contributions, cooperation, and patience; and to Cliff Mills, and his predecessor, Joyce Berry, at Oxford University Press, for their personal interest in this project. Their belief in this cross-disciplinary approach and their insightful comments enhanced the clarity of the presentation. We are especially grateful to Charlotte Hitchcock of New Haven, CT, for the thoroughness and care she provided in the design and preparation of the manuscript for e-transcription.

Contents

Part I: Natural Geologic Hazards

A. Physical — Obvious and Immediate

B. Chemical— "Silent" and Long Term

Part II Anthropogenic Changes to the Geologic Environment

Part III Identifying the Hazards

List of Contributors

Note: Only the lead authors show email addressses.

Antony R. Berger
Victoria, British Columbia, Canada
aberger@uvic.ca

Peter W. Abrahams
Institute of Geography and Earth Sciences
University of Wales, Aberystwyth SY23 3DB, U.K.
pwa@aber.ac.uk

Harvey E. Belkin
U.S. Geological Survey, MS 956, Reston, VA 20192

B. Carlmark
Centre for Metal Biology in Uppsala
Rudbeck Laboratory, SE-751 85 Uppsala, Sweden

Jose A. Centeno
Armed Forces Institute of Pathology
Washington, D.C. 20306

Portia O. Ceruti
University of Stellenbosch
Department of Soil Science
Private Bag X1, Matieland 7602, South Africa
poceruti@maties.sun.ac.za

Alan Condron
Institute of Geography and Earth Sciences
The University of Wales Aberystwyth, SY23 3DB, U.K.

Sabrina Coty
Chemistry Department, Xavier University of Louisiana
1 Drexel Drive, New Orleans, LA, 70125

J.G. Crock
U.S. Geological Survey, MS 973
Denver Federal Center, Denver, CO 80225

A. Danersund
Centre for Metal Biology in Uppsala
Rudbeck Laboratory, SE-751 85 Uppsala, Sweden

T.C. Davies
School of Environmental Studies, Moi University
P.O. Box 3900, Eldoret, Kenya
tcdavies@net2000ke.com

Devra L. Davis
Carnegie Mellon University
Heinz School, Pittsburgh, PA 15213

W.C. Day
U.S. Geological Survey, MS 973
Denver Federal Center, Denver, CO 80225

Edward Derbyshire
Centre for Quaternary Research
Department of Geography, Royal Holloway
University of London, Egham, Surrey TW20 0EX, U. K.
100666.1577@compuserve.com

Dorina Diaconita
Geological Institute of Romania
Caransebes Str. 1, RO-78344, Bucharest, Romania

Chandra B. Dissanayake
Dept. of Geology, University of Peradenyia
Kandy, Sri Lanka

Victor Dumitrascu
Clinical Laboratory No.1, County Hospital Timisoara
Str. L. Rebreanu 156, RO-1900 Timisoara, Romania

Michael Durand
Natural Hazards Research Centre, Dept. of Geological Sciences
University of Canterbury, Private Bag 4800, Christchurch, N.Z.

Gerald L. Feder
Florida Community College at Jacksonville
Jacksonville, FL 32256

Martin Fey
University of Stellenbosch, Department of Soil Science
Private Bag X1, Matieland 7602, South Africa

Robert B. Finkelman
U.S. Geological Survey, MS 956, Reston, VA 20192
rbf@USGS.gov

Fiona M. Fordyce
British Geological Survey, West Mains Road
Edinburgh, EH9 3LA, U. K.
fmf@bgs.ac.uk

Adrian Frank
Department of Clinical Chemistry
Faculty of Veterinary Medicine,
Swedish University of Agricultural Science
Box 7038, SE-750 07 Uppsala, Sweden
Dr.a.frank@Rocketmail.com

David Gilbertson
School of Conservation Sciences
Bournemouth University, Dorset BH1 5BB, U. K.

C. Gonzales
College of Pharmacy, Xavier University of Louisiana
1 Drexel Drive, New Orleans, LA, 70125

L.P. Gough
U.S. Geological Survey, 4200 University Dr., Anchorage, AK 99508
lgough@usgs.gov

John Grattan
Institute of Geography and Earth Sciences
University of Wales Aberystwyth SY23 3 DB, U.K.
jpg@aber.ac.uk

S-O. Grönquist
Centre for Metal Biology in Uppsala
Rudbeck Laboratory, SE-751 85 Uppsala, Sweden

Michalann Harthill
U.S. Geological Survey, MS 300, Reston VA 20192
michalann_harthill@usgs.gov

Gunnar Hillerdal
Departments of Lung Diseases
Karolinska Hospital, S-17176 Stockholm, and
      Akademic Hospital, Uppsala, Sweden
gunnar.hillerdal@lung.aus.lul.se

Hou Shaofan
Institute of Geographical Sciences and Natural
      Resources Research
Chinese Academy of Sciences
Beijing 100101, P. R. China

R. Hudecek
Centre for Metal Biology in Uppsala
Rudbeck Laboratory, SE-751 85 Uppsala, Sweden

Chris C. Johnson
British Geological Survey, Keyworth
Nottingham, NG12 5GG, U. K.

Lotus Abu Karaki
Department of Antiquities, Amman, Jordan

U. Lindh
Centre for Metal Biology in Uppsala
Rudbeck Laboratory, SE-751 85 Uppsala, Sweden
ulf.lindh@bms.uu.se

A. Lindvall
Centre for Metal Biology in Uppsala
Rudbeck Laboratory, SE-751 85 Uppsala, Sweden
anders.lindvall@spray.se

Susan V. M. Maharaj
US Geological Survey, MS 956, Reston, VA 20192

H.W. Mielke
College of Pharmacy
Xavier University of Louisiana
1 Drexel Drive, New Orleans, LA, 70125
hmielke@xula.edu

P. R. Nadebaum
CSIRO Land and Water
PMB No. 2, Glen Osmond
Adelaide, South Australia 5064, Australia

R. Naidu
CSIRO Land and Water
PMB No. 2, Glen Osmond
Adelaide, South Australia 5064, Australia
ravi.naidu@adl.clw.csiro.au

Udaya R. B. Navaratne
Dept. of Geology, University of Peradenyia
Kandy, Sri Lanka

Colin Neal
Centre for Ecology and Hydrology
Maclean Building, Crowmarsh Gifford
Wallingford, Oxfordshire, OX10 8BB, U. K.
cn@ceh.ac.uk

William H. Orem
US Geological Survey, MS 956, Reston, VA 20192

Virgil Paunescu
Clinical Laboratory No.1, County Hospital Timisoara
Str. L. Rebreanu 156, RO-1900 Timisoara, Romania

Maciej Pawlikowski
Laboratory of Biomineralogy
Institute of Mineralogy, Petrography and Geochemistry,
University of Mining and Metallurgy
al. Mickiewicza 30, 30-059 Krakow, Poland
mpawlik@uci.agh.edu.pl

Jane A. Plant
British Geological Survey
Keyworth, Nottingham NG12 5GG, U.K.
japl@bgs.ac.uk

Justin Pooley
University of Stellenbosch, Department of Soil Science
Private Bag X1, Matieland 7602, South Africa

E. Powell
College of Pharmacy, Xavier University of Louisiana
1 Drexel Drive, New Orleans, LA, 70125

F. Brian Pyatt
Department of Life Sciences, Nottingham Trent University
Clifton Lane, Nottingham NG11 8NS, U. K.

Shaun Reeder
British Geological Survey
Keyworth, Nottingham NG12 5GG , U.K.

Eleanora I. Robbins
San Diego State University
Dept. Geological Sciences MC-1020
5500 Campanile Dr., San Diego, CA 92182-1020
nrobbins@geology.sdsu.edu

Ziad al Saad
The Institute of Archaeology and Anthropology
Yarmouk University, Irbid, Jordan

O. Selinus
Geological Survey of Sweden
PO Box 670, SE-75128 Uppsala, Sweden
olle.selinus@home.se

Aila Shah
Chemistry Department, Xavier University of Louisiana
1 Drexel Drive, New Orleans, LA, 70125

H. Catherine W. Skinner
Department of Geology and Geophysics
Yale University, Box 208109, New Haven, CT 06520
catherine.skinner@yale.edu

Barry Smith
British Geological Survey
Keyworth, Nottingham NG12 5GG, U.K.

David Smith
US Geological Survey, MS 973
Denver Federal Center, Denver, CO 80225
dsmith@usgs.gov

Eiliv Steinnes
Department of Chemistry
Norwegian University of Science and Technology
N-7491 Trondheim, Norway
Eiliv.steinnes@chembio.ninu.no

Diana N. Szilagyi
Clinical Laboratory No.1
County Hospital Timisoara
Str. L. Rebreanu 156, RO-1900 Timisoara, Romania

Tan Jian'an
Institute of Geographical Sciences and
    Natural Resources Research
Chinese Academy of Sciences
Beijing 100101, P. R. China

Calin A. Tatu
Clinical Laboratory No.1, County Hospital Timisoara
Str. L. Rebreanu 156, RO-1900 Timisoara, Romania
cta@med.unc.edu, ctatu@usgs.gov

Sharon Taylor
Institute of Geography and Earth Sciences
The University of Wales Aberystwyth, SY23 3DB, U.K.

Wang Wuyi
Institute of Geographical Sciences and
    Natural Resources Research
Chinese Academy of Sciences
Beijing 100101, P. R. China
wangwy@igsnrr.cn

Yang Linsheng
Institute of Geographical Sciences and
    Natural Resources Research
Chinese Academy of Sciences
Beijing 100101, P. R. China

Zheng Baoshan
Institute of Geochemistry
Chinese Academy of Sciences
Guiyang, Guizhou Province, P. R. China 550002

Introduction

This volume is a contribution to the new and rapidly expanding field of medical geology that links geologists and other earth scientists with plant and animal biologists and medical, dental, and veterinary specialists in efforts to resolve local and global health issues. The topics mentioned range from the health effects of arsenic, mercury, and fibrous minerals, natural hazards that contribute to the etiology of endemic diseases, to questions on the identification of such hazards. Medical geology aims to strengthen and integrate research that can reduce environmental threats to the health and well-being of humans and animals. It embraces disciplines as diverse as mineralogy and pathology (Geology and Health 2001, Geosciences and Human Health 2001).

Health generally refers to people and other living creatures, whereas the focus of geology is on the inanimate and the distant past. Although these may be separate arenas or compartments for investigations, the direct links are hard to ignore. Life itself has evolved within a matrix of earth materials — rocks, minerals, soils, water, air — the availability of which has a profound control on what all living creatures ingest and how they develop, both biologically and culturally. The air we breathe, the water we drink, and the nutrients we consume depend on the geological environment that we can only partially control. As we struggle to cope in a world rushing toward 10 billion people, a better understanding of the ways in which the natural environment influences our health should permit more intelligent decisions for the future. The general consensus concerning global change recognizes that humans have had a powerful impact on their surroundings. The other side to that relationship — the sometimes harmful effects of geological materials and processes on us — is the subject of this volume.

Combining knowledge and expertise from the earth sciences with that from the medical and life sciences has numerous applications to the resolution of health issues. Coordinating efforts can sharpen the definition of a problem, aid in strategies of reclamation, define and locate sources of potable water, and develop economical solutions based on geological principles that can help to ease, if not prevent, suffering and disease. All of these possibilities are discussed in this book, justifying the conclusion that geoscientists will continue to contribute to the improvement of global public health.

This volume grew out of a meeting held in Sweden in September 2000, where 40 scientists — medical, dental, veterinary, and geological — presented their work on the relationships between human and animal health and rocks, soils, and water. Hosted at the Swedish Geological Survey in Uppsala by Dr. Olle Selinus, the seminar on "Health and the Geological Environment" provided a forum for discussions of well-known hazards, such as dust and arsenic, and some of the nutrients required for a sustainable and healthy existence, together with their sources, and transport mechanisms. There was free and lively exchange among participants from Europe, Asia, Australia, Africa, and North America, and the topics ranged from the submicroscopic to the global levels.

The topics presented here are a kind of reconnaissance, a sampling, of the spectrum of research now taking place in the broad space between geology and medicine. The chapters are neither fully developed statements of research efforts nor comprehensive treatments of specific issues, but rather they are brief pictures of various aspects of the crossover between geology and health. The wide range of authors, their individual expertise, geographic distribution, and different writing styles reflect the diversity of the medical geology "project," each presentation reflecting in part the approaches common in widely differing specialist fields. Some contributions are highly focused on local or small-scale problems, while others deal with very broad issues. Some are almost telegraphic in their brevity, while others are more discursive. Some can be readily understood by nonspecialists, while others are more technical. As editors, we have attempted to bring some coherence to this diversity by a series of commentaries, which precede and set the scene for the chapters that follow Berger's introduction. These we have arranged in three groups.

The **first** group deals with natural geological (geogenic, to use Naidu and Nadebaum's term) hazards, both physical and chemical. The physical hazards are obvious and immediate, and are commonly difficult or impossible to avoid. All of the remaining chapters in this section focus on natural geochemical hazards, which are "silent" and long term. A great deal of information is available on many of the 92 naturally occurring elements, defining their chemistry, bio- and geochemistry, and some are well known, but their geological distribution and potential impacts have only fairly recently come to the fore. The **second** part of this volume deals with some of the ways human actions on the geological environment can affect health. These contributions focus on the composition and distribution of the chemical species in soils, rocks, and water presently known using today's technology and sensitive methods of detection. These alterations of the geoenvironment as the result of human activities may affect health in ways we have only begun to appreciate. The chapters in the **third** group illustrate how health hazards can or have been identified and quantified. A **glossary** that covers some of the technical medical terms is included, together with a list of sources consulted.

It is our intention to illuminate both the diversity of potential problems and the possibilities for crossovers between geology and health by juxtaposing contributions from different geographic and specialist areas, examples from the myriad of opportunities that exist in the continuum between the biosphere, the lithosphere, the hydrosphere, and the atmosphere. What we present here is only a taste of the challenges, the tip of the iceberg. We regret that there are no contributions in this collection specifically devoted to plants or to agricultural issues. These are often of critical concern because the transport of nutrients and of hazards, from rocks and soils to living creatures, identifies plants as essential sources and pathways. It is our hope that readers will gain fresh insights from this collection and that the presentations will stimulate new joint efforts between scientists and health practitioners.

References

Geology and Health (2001) Issue 17 of Earthwise, British Geological Survey, Keyworth, Nottingham, NG12 5GG.

Geosciences & Human Health (2001) Vol. 46 #11 of Geotimes, American Geological Institute, 4220 King St. Alexandria, VA 22302

July 2002

H. Catherine W. Skinner
New Haven, Connecticut

Antony R. Berger
Victoria, British Columbia

1

Linking Health to Geology

Antony R. Berger

What Is Medical Geology?

In staking the ground for any new field of science, its distinct character needs to be established. In our opinion, the already large literature on geology and health, including the chapters in this volume, provide two clear arguments for distinctiveness. **First**, medical geology extends the primary concern of geologists with the interactions between rocks, soils, water, and air to the effects of these interactions on the health of humans and other living organisms. Though one focus of medical geology is the search for the origins of disease in the natural geological background, there is also interest in the obvious benefits that the major, minor, and trace elements and the essential molecules found in soils, surface, and groundwater, and in the air we breathe, bring to health and well-being. **Second**, this new field is truly cross-disciplinary; it requires the melding of two distinct research efforts, the one focused on geology, with all its subdisciplines, and the other on living forms. Different viewpoints can be myopic, and to increase understanding of the health implications of the natural background requires the involvement not only of a wide range of earth scientists, but also of researchers and practitioners in medicine, dentistry, veterinary science, biology, botany, agriculture, and ecology, among others.

From the viewpoint of the life scientists, medical geology could be regarded as a subdivision of "environmental medicine" (Möller 2000). This increasingly important aspect of medicine includes consideration of airborne pathways of disease, ozone depletion, algal blooms, the organohalogens, and mycotoxins found as part of the 'ecology' of the built environment (buildings, factories). In general, the purview is any factor in the natural or human environment that affects health. The term "geomedicine" has been used extensively, especially by the late J. Låg (1990). However, unlike the well-established fields of geophysics and geochemistry, in which physics and chemistry are applied to geology, the new field is clearly not about the relevance of medical principles to geology. Rather, it is concerned with the application of geological knowledge and techniques to a more integrated approach to public health. Moreover, as pointed out in the Uppsala meeting, geomedicine seems unlikely to gain acceptance among medical professionals, who use a range of adjectives and modifiers to indicate specialities such as orthopedic medicine, nuclear medicine, and family medicine. The popular current alternative name is "medical geology," as in engineering geology, Quaternary geology, or even military geology.

As long as the new field is being driven from the earth sciences, medical geology seems to fit the bill well enough. It parallels the well-established field of medical geography, which includes the geographic factors of virtually any disease or public health issue (Meade and Earickson 2000). Whatever the final name chosen, medical geology can be described as "the study of the linkages between natural geological factors and the health and well-being of humans and other living organisms." Note the inclusion of nonhuman portions of the biosphere whose health is essential to human survival. This is a slightly modified version of the definition being used by IGCP Project 454 and the IUGS Working Group on Medical Geology. Although it excludes plants and animals, a somewhat more concise explanation is that medical geology is the study of the "public health impacts of geological materials and processes." According to this view, medical geology includes "identifying and characterizing natural and anthropogenic sources of harmful materials in the environment, learning how to predict the movement and alteration of chemical, infectious, and other disease-causing agents over time and space, and understanding how people are exposed to such materials and what can be done to minimize or prevent such exposure" (see <http://www.ecosystemhealth.com/hehp/groups.htm>).

On Crossing Over Discipline Boundaries

This book extols the virtues of cross-disciplinary research in reducing environmental threats to health and well-being. E.O. Wilson (1998a) discusses some of the ground to be explored between the natural and social sciences, using the term 'consilience' to refer to "the interlocking of causal explanations across disciplines." His argument can perhaps be extended to view the boundary between medicine and geology, with all their subfields, not as a wall, but as "a broad, mostly unexplored domain of causally linked phenomena" (ibid. p. 2048; see also Wilson 1998b). For Wilson, the central question of social science is the linkage between genetic and cultural evolution. There is an analogy in the linkage between geology and medicine. Since what we ingest has surely been a major factor in how we have evolved, and since food and drink, which are basic to culture, have as their ultimate source the earth, its materials and processes, then one may expect that crossing over from geology to medicine will contribute to understanding this linkage. Seeing ourselves, like other species, as part of the geological continuum may provide insights about our relationship with the environment, and in turn influence the way we act towards the nonhuman, natural world. An exploration of the medicine-geology divide may help to reconceptualize both the beneficial and the harmful sides of the natural world and its processes, provide a better understanding of the "physical basis of human nature, down to its evolutionary roots and genetic biases" (Wilson 1998a, p. 2049), and yield insights on ways to diagnose and manage some of the global crises.

Frodeman et al. (2001) point out that there is now no field of study that takes as its origin understanding the relationship between the disciplines. Although new disciplines commonly spring from interdisciplinary ventures, once formed they are themselves liable to set up new discipline boundaries, which can in turn block further crossovers. They argue that crossover studies can also advance the different disciplines involved. To most people, medicine and geology seem very odd bedfellows, the former dealing with healing people, the latter with rocks and minerals and the distant past. Could it be that exploring the wide gap between them might bring new insights into what counts as knowledge?

If so, the development in the coming years of this very diverse field might provide an epistemological experiment well worth following. Perhaps some future student of consilience will turn back to the beginnings of the twenty-first century for an example of a new interdisciplinary field here striving to be born.

But there is also a dilemma, for interdisciplinarity is very much a two-way street. Yet, medical geology is being driven at the outset mainly from the earth sciences, and especially from geochemistry. The cry is "If only the medical experts would listen to us, we could resolve many serious problems." Though it is becoming somewhat more common today for geoscientists, epidemiologists, ecologists, and veterinary researchers to cooperate in identifying causes of diseases, it has not been easy to attract medical researchers to cooperative efforts. Medicine concentrates first and foremost on relieving pain and suffering, so that it is necessary to demonstrate how geological information can benefit patients. The review of medical geology by Selinus and Frank (2000) provides many examples of how this is being done, and the present volume also contributes to this understanding.

Why Has It Taken So Long?

If the relationship between rocks, water, soils, and health is so important, and if its antecedents are so ancient — natural salt for human consumption goes back into the dim recesses of history — why hasn't the crossover advanced further? A partial answer may lie in the organizational barriers to interdisciplinary research. Within the earth sciences, linkages are generally strong between many subfields, such as hydrology and hydrogeology, for there are few sharp boundaries between the behavior of surface and subsurface water. However, soils are generally regarded by geologists as within the purview of agriculture and, thus, outside their range of interests. The traditional focus in pedology has been on the descriptive side of soils, such as taxonomy, soil profiles, and catenas. However, the present focus on environment and health issues reinforces the importance of soil as the membrane between rocks and most living organisms, comparable in a sense to the walls of cells, which act as gatekeepers for signals from the rest of the body.

The many separate societies with competing research projects and goals within the geoscience community, while providing for organizational diversity, may also act as a barrier to working across disciplinary boundaries. For example, the International Association of Geochemistry and Cosmochemistry has a working group on medical geology. There is a long-established Society for Environmental Geochemistry and Health, and several other international earth science groups have activities related to health issues, including the International Association for Volcanology and Chemistry of the Earth's Interior, and the International Association of Hydrological Sciences. Oddly enough, the very large and international Society for Environmental Toxicity and Chemistry, though dealing with many of the concerns of medical geology, does not list geology or any of its major subfields among the many disciplines it embraces. Until recently the major agencies funding scientific research have made crossing from one discipline to another rather difficult, though the National Science Foundation (U.S.) and the National Research Council (Canada) have recently signaled the importance of multidisciplinary research.

Are Silent Geochemical Threats Natural Hazards?

There can be no doubt that certain naturally occurring materials, elements, and chemicals can be harmful to human, animal, and plant health: toxic elements in groundwater, the "poisoning" effect of natural heavy metals on vegetation, or deficiencies of essential elements in food crops. The spatial association between health hazards and geochemical "anomalies" that reflect certain rock types and mineral deposits is wellknown, as in the case of radon (Brookins 1990). In Norway, where authorities have tried to encourage greater use of groundwater as a cheap, high-quality alternative to surface water, more than 50% of all bedrock wells exceed recommended drinking water limits for one or more health-related parameters, particularly radon and fluoride (Banks et al. 1998). Hazards to health in all these cases stem from natural conditions that may be pervasive and temporally continuous, or that may vary in time and space with changes in climate, erosion patterns, volcanicity, or the chemical composition of air or surface and groundwater.

It is, thus, rather surprising that the very active and interdisciplinary study of natural hazards, which has its own extensive scientific and policy literature and many national and international agencies and programs, generally does not seem to include consideration of dangers to health from non-anthropogenic chemicals and reactions in soils, water, and air. Such geogenic hazards are well illustrated in the extensive literature on environmental geochemistry (Appleton et al. 1996, Bowie and Thornton 1985). The UN Disaster Relief Office has defined natural hazards broadly as "the probability of occurrence within a specified period of time and within a given area of a potentially damaging phenomenon" (UNDRO 1982), and a number of publications also include hazards from geological materials (e.g. Nuhfer et al. 1993, Guo 1991). Yet, for the most part, the literature on natural hazards ignores the kind of "silent" threats to health that are standard components of the natural environment (Bryant 1991, Burton et al. 1993, Chapman 1994, Tobin and Montz 1997). Even some of the key books on medical geography and epidemiology have very little discussion of the actual or potential effect on health of natural geochemical hazards in rocks, soils, and water (e.g. Howe 1997, Meade and Earickson 2000). Perhaps it is the temporal dimension of geochemical hazards that has kept them distinct from the standard literature on natural hazards, which most hazards workers link to "events," those with obvious beginnings, middles, and ends, and usually quite short — less than a season, a week or a day. Yet those who include biological threats, such as malaria and schistosomiasis (Chapman 1994), in their discussion of natural hazards, clearly extend the temporal perspective to essentially continuous hazards.

The bulk of the literature on natural hazards discusses them solely in terms of their effect on humans and society. As Gilbert White, the pioneer researcher in this field put it, "by definition, no natural hazard exists apart from human adjustment to it" (1974, p.3). For Burton and Kates (1964), natural hazards are those components of "the physical environment harmful to man and caused by forces extraneous to him." Doornkamp (1989) defines a geohazard as one of a "geological, hydrological or geomorphological nature, which poses a threat to man and his activities." This anthropocentric position seems to be echoed by many contemporary writers on natural hazards and disasters. But from a biospheric perspective that recognizes harm done to organisms and even whole ecosystems, the scope of natural hazards extends well beyond human society. There is probably no greater hazard than that caused by the impact of a comet or asteroid, of the sort that hit the Earth at various times in the distant geological past, resetting the evolutionary clock. At the other end of the scale there are numerous examples of the harmful effects of toxic elements in soils and water on animals and fish (e.g., McCall et al. 1966). The ecological (and the mineral exploration) literature describes many situations where plant growth has been retarded by the presence of various chemicals that inhibit growth, structure, or normal plant functions, and where plants act as accumulators of elements harmful to grazing animals. Medical geology extends concerns to the health of animals and vegetation, and thus may help to balance anthropocentric attitudes toward the environment and to stress continuities and crossovers within the biosphere.

Medical Geology and Environmental Management Issues

Turning now to practical applications, medical geology can help to develop guidelines for managing environments where high or low natural background levels of elements, compounds, and minerals in rocks, soils, and water pose a threat to human, plant, or animal health. In doing so, it might be helpful to develop a simple taxonomy of geomedical hazards, as has been done through the geoindicator approach (Berger and Iams 1996) for rapid and dynamic geological processes, some of which can also affect health. These include:

- Increases in sediment erosion and transport, which can uncover toxic wastes deposited earlier ("chemical time bombs")
- Permafrost and gas-hydrate melting, which can lead to methane degassing and influx of insects and diseases
- Changes in groundwater and surface water quality resulting from accelerated erosion, deforestation, pollution, river diversion, groundwater mining
- Drop in lake levels and increase in salinity causing loss of water supplies and increase in airborne particulates
- Catastrophic natural events — landslides and avalanches, earthquakes, tsunamis, and volcanic eruptions, with many effects on settlements, water, soils, air quality

Such a classification could be valuable to environmental planners and the general public, especially if it identified the relevant pathways from source to organism, and distinguished harmful chemicals of natural origin from those due to human actions such as industrialization, urbanization, and waste disposal. If this is not done, serious policy issues can arise.

For example, artisanal gold miners have been criticized severely for being careless with mercury, and there have been many efforts to reduce its usage and to prevent its loss to the environment where it constitutes a health hazard. However, geochemical research shows that much of the mercury around placer gold mining operations in the Tapajós basin of Brazil (Telmer et al. 2000) and in the Rio Béni basin of Bolivia (Maurice-Bourgoin et al. 1999) is bound to suspended sediments eroded from gold-bearing soils that were high in natural Hg long before the advent of humans. Reducing mercury pollution of the rivers and streams would thus require a reduction of placer operations, and not merely a decrease in the use of Hg in gold recovery. The importance of understanding the pathways taken by natural contaminants is clear.

The general public seems to take the "anthropoblamist" view that high concentrations of harmful metals and chemicals must be the result of human activities such as industry, energy generation, and waste disposal. In Canada, for example, a common opinion is that metals in the soils and waters of remote areas come from the long-range atmospheric transport of particulates emitted from smelters, power plants, and waste incinerators. Yet detailed studies of the geochemistry of waters, soils, and bedrocks show that elevated levels of some potentially toxic elements are due to local geological conditions and were present before anthropogenic disturbance of the environment (Rasmussen et al. 1998). Of course, human activities frequently induce changes in natural geochemical backgrounds. Soil pH in parts of Finland has decreased from neutral to 3–4 within 40 to 60 years as a result of reforestation of formerly agricultural fields, with the result that heavy metals like Co, Cr, and Cu, have become more bioavailable and harmful to biota (Lahdenpera 1999). Thirty years of careful monitoring in Finland have shown marked changes in groundwater quality, especially as a result of acidification in the 1970s and 1980s from both natural and human inputs (Backman et al. 1999).

These broad considerations emphasize the need to understand the linkages between the natural geological background and health in order to gain a proper understanding of geochemical cycling and the natural phenomena that contribute to it. A major challenge is to find ways to distinguish between natural inputs, which may not be manageable, and human-induced ones that are. The emerging field of medical geology is poised to make an important contribution to this effort.

References

Appleton, J.D., R. Fuge and G.J.H. MCCall (eds) 1996. Environmental geochemistry and health with special reference to developing countries. London: Geological Society, Special Publication 113, 264p.

Backman, B., P. Lahermo, U. Vaisanen, T. Paukola, R. Juntunen, J. Karhu, A. Pullinen, H. Raino & H. Tanskanen 1999. The effects of geological environment and human activities on groundwater in Finland. The results of monitoring in 1969–1996. Geological Survey of Finland Report of Investigations 147, 155–167.

Banks, D., A.K. Midtgard, G. Morland, C. Raimnan, T. Strand, K. Bjorvatn & U. Siewers 1998. Is Pure Groundwater Safe to Drink?: Natural Contamination of Groundwater in Norway. Geology Today, May–June, 104–113.

Berger A.R. & W.J. Iams 1996. Geoindicators. Assessing rapid environmental changes in earth systems. Rotterdam: A.A. Balkema, 466p.

Bowie, S.H.U. and I. Thornton (eds) 1985. Environmental geochemistry and health. Boston: D. Reidel, 140p.

Brookins, D.G. 1990. The indoor radon problem. New York, Columbia University Press, 229p.

Bryant, E. 1991. Natural hazards. Cambridge: Cambridge University Press, 294p.

Burton, I. & R.W. Kates 1964. The perception of natural hazards in resource management. Natural Resources Journal vol. 3, 412–414

Burton , I., R.W. Kates and G.F. White. 1993. The environment as hazard (Second edition). New York: The Guilford Press, 293p.

Chapman, D. 1994. Natural hazards. Melbourne: Oxford University Press, 174p.

Doornkamp, J.C. 1989. Hazards. In G.J.H. McCall & B.R. Marker (eds.). Earth science mapping for planning, development and conservation. London: Graham and Trotman, p157–173.

Frodeman, R., C. Mitcham & A.B. Sachs 2001. Questioning interdisciplinarity. Science, Technology and Society Newsletter, Nos. 126–127, p1–5.

Guo Xizhe (ed) 1991. Geological hazards in China and their prevention and control. Beijing: Geological Publishing House, 259p

Howe, G.M. 1997. People, environment, disease and death. A medical geography of Britain throughout the ages. Cardiff: University of Wales Press, 328p.

Lahdenperä, A-M. 1999. Geochemistry of afforested and arable soils in Finland. Geological Survey of Finland, Current Research, Special Paper 27, p61–68.

Maurice-Bourgoin, L., I Quiroga, J-L. Guyot & O. Malm. 1999. Mercury pollution in the upper Béni River, Amazonian Basin, Bolivia. Ambio vol 28 (4), 98–101.

Meade, M.S. and R.J. Earickson 2000. Medical geography (second edition). New York: The Guilford Press, 500p.

Nuhfer, E.B., R.J. Proctor & P.H. Moser 1993. The citizens' guide to geologic hazards. Arvada CO: American Institute of Professional Geologists, 134p.

Rasmussen, E., P.W.B. Friske, L.M. Azzaria, & R.G. Garrett 1998. Mercury in the Canadian environment: current research challenges. Geoscience Canada, 25 (1) 1–13.

Selinus, O. & A. Frank 2000. Medical geology. In L. Möller (ed.). Environmental medicine. Stockholm: Joint Industrial Safety Council, 164–183.

Telmer, K, M. Costa, E.S. Araujo, R.S. Angelica and Y. Maurice, 2000. Halting and repairing contamination – the importance of understanding source: alluvial gold mining and mercury pollution in the Tapajos River basin, Para, Brazilian Amazon, a case study. COGEOENVIRONMENT Newsletter 17, 8–10.

Tobin, G.A. & B.E. Montz 1997. Natural hazards. Explanation and integration. New York: The Guilford Press, 388p.

UNDRO 1982. Natural disasters and vulnerability analysis. Office of the UN Disaster Relief Coordinator, Geneva, 76p.

White, G.F. (ed.) 1974. Natural hazards: local, national, global. NY Oxford University Press, 267p.

Wilson, E.O. 1998a. Integrated Science and the Coming Century of the Environment. Science, Vol 279 (5359), pp. 2048–2049.

Wilson, E.O. 1998b. Consilience. The unity of knowledge. New York: A.A. Knopf, 332p.

Part 1

Natural Geologic Hazards

A. Physical — Obvious and Immediate

The first section of this volume deals with naturally occurring hazards to human and animal health, borrowing the term 'geogenic' used by **Naidu and Nadebaum.*** It is not our intention to add to the voluminous literature on well-known geological hazards to health associated with landslides, floods, volcanic eruptions, and other natural physical phenomena. These are widely reported in the media and analyzed in the scientific press. Instead the following two chapters discuss some of the less well-known dangers associated with geological processes.

Dust and mineral aerosols are now recognized as major components of the tropospheric aerosol assemblage, with the global annual input to the atmosphere estimated at 1–2 billion metric tons/year (Bergametti and Dulac 1998). Dust raised from the central Asian deserts has been detected 5000 km away, and particulates from the Sahara are regularly seen on the western side of the tropical and semi-tropical Atlantic. There are many data available on the distribution of dust in the atmosphere and the chemical and physical characteristics of the particulates (Prospero 2001). Dust and mineral aerosols may provide iron and other micronutrients to marine and terrestrial ecosystems, but where concentrated and transported by huge windstorms, they also present an obvious physical hazard to human health. Though most arid regions with drifting sands are sparsely inhabited, there are areas where established populations have endured for centuries in dusty environments. One is northern China where, as **Derbyshire** shows, the vast loess deposits, themselves of aeolian origin, can pose a serious health hazard. Particulates, mobilized and transported by wind, may be carried as far away as NW India and probably even farther. Little seems to be known in China about the health hazards of airborne loess. A proper analysis of such risks will require better understanding of wind and accumulation patterns, and the incidence of dust-related diseases, such as silicosis, with the possible confounding links such as smoking habits, information that is scarce for most of the Chinese population.

Using sensitive modern analytical techniques, researchers are documenting the diseases caused by wind-blown mineral and biogenic dusts. One locally researched disorder known as Valley Fever has been investigated in California (Jibson et al. 1998). On a regional scale,

* *References to chapters in this volume are in* **bold***.*

13

dust storms whipped up from the dried bed of the Aral Sea transport a range of dangerous aerosols, some of natural origin, but many related to pesticides earlier applied to agricultural fields in the area. There is extensive information about the environmental and agricultural consequences of the Aral Sea "tragedy," and the people in the region suffer many dire health problems, though there appears to be less known about the way these are related to dust (<www.grida.no/aral/aralsea>).

Harm to health caused by globally transported airborne particulates and gases is not a new phenomenon (McCoy and Heiken 2000). **Grattan, Taylor, et al.** attribute unusually high death rates in rural England in the summer of 1783 to the eruption of Laki in Iceland, an effect ascribed to far-transported aerosols, or volcanic fogs, high in SO_4. Aerosols emitted from volcanoes today are tracked by satellite and remote sensing instruments. Such global observations could be used to warn people to take precautions or even to move out of harm's way. The prevention of health hazards from dust provides an illustration of coupling geologic and medical information.

References

Bergametti, G. and F. Dulac. 1998. Mineral aerosols: renewed interest for climate forcing and tropospheric chemistry studies. Global Change Newsletter, v.33, p.19–23.

Jibson, R.W., Harp, E.L., Schneider, E., Hajjeh, R.A., and Spiegel, R.A., 1998, An outbreak of coccidioidomycosis (valley fever) caused by landslides triggered by the 1994 Northridge, California, earthquake: Geological Society of America Reviews in Engineering geology, Volume XII, p. 53–61.

McCoy, Floyd W. and Heiken, Grant (2000) Volcanic hazards and disasters in human antiquity. Special Paper #345, Geological Society of America, Boulder, CO.

Prospero, Joseph M. (2001) African dust in America. Geotimes v.46 (11), p.24–27.

2

Natural Dust and Pneumoconiosis in High Asia

EDWARD DERBYSHIRE

Centre for Quaternary Research, Department of Geography
Royal Holloway, University of London, Egham, Surrey, United Kingdom

High Asia, defined here as that great tract of land from the Himalaya-Karakoram in the south to the Tian Shan in the north and the Pamir in the west to the Qinling Mountains in the east, is a very dusty place. Whole communities of people in this region are exposed to the adverse effects of natural (aerosolic) dusts at exposure levels reaching those encountered in some high-risk industries. Outdooor workers are at particular risk. However, few data are available on the magnitude of the dust impact on human health. The effect of such far-travelled particles on the health of the human population in the Loess Plateau, and including major Chinese cities, has received relatively little attention to date.

A combination of the highest known uplift rates, rapid river incision (up to 12 mm/yr: Burbank et al. 1996), unstable slopes, glaciation and widespread rock breakup by crystal growth during freezing (frost action), and by hydration of salts (salt weathering) makes the High Asia region the world's most efficient producer of silty (defined as between 2 and 63 μm) debris.

The earliest written records of the dust hazard come from China, most notably in the "Yu Gong" by Gu Ban (ca 200 BC) (Wang and Song 1983). Here, deposits of wind-blown silt (known as 'loess') cover the landscape in a drape that is locally 500 m thick. In North China, the loess covers an area of over 600,000 km², most of it in the Loess Plateau, situated in the middle reaches of the Huang He (Yellow River).

The characteristic properties of loess include high porosity and collapsibility on wetting (Derbyshire et al. 1995, Derbyshire and Meng 2000). Thus, it is readily reworked and redistributed by water. This process concentrates silts in large alluvial fans (up to 50 x 50 km) in the piedmont zones of 6,000 m high glacier- and snow-covered mountain ranges of western China, including the Altai Shan ('shan' = mountains), Tian Shan, Kunlun Shan, Qilian Shan, and Karakoram. These zones are loci for human populations, and also a major source of wind-blown dust.

The Chinese loess has a simple mineralogy consisting of angular, blade-shaped quartz grains (~60–65%), with minor feldspars, micas, and carbonates. Clay minerals rarely total more than 12–15% in the loess units, although they may reach 40% in the intercalated soils (palaeosols). The air-fall nature of deposition results in high to very high porosities and low bulk densities (> 40% and ~1400 kg/m³, respectively, in the most recent loess). Considerable convergence of cold and dry westerly airflows leeward of the Tibetan Plateau, driven by the Mongolian 'high' and orographically enhanced, is characteristic of the winter/spring circulation pattern. The 30-year mean annual incidence of dust storm days exceeds 35 in the Taklamakan Desert, the Hexi Corridor, and the western Loess Plateau (Fig. 2.1). Substantial dust falls (0.02–0.08 mm/yr, equivalent to ~32–108 tons/km²/ yr) occur in at least 6 months of the year at several sites in this region (Derbyshire et al. 1998).

15

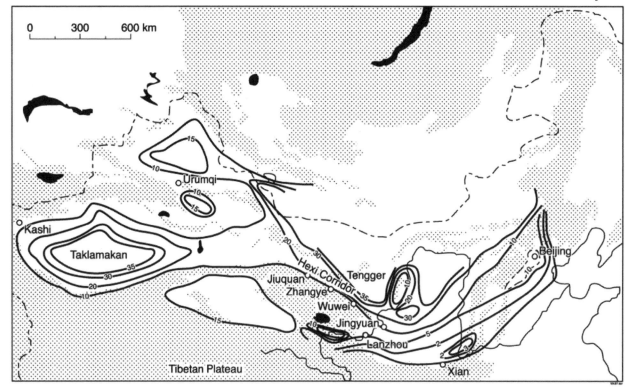

Figure 2.1: Annual number of sand and dust storm days (30–year mean) in North and West China. Shaded areas are mountains and plateaux. Lakes are shown in black (after Derbyshire et al. 1998).

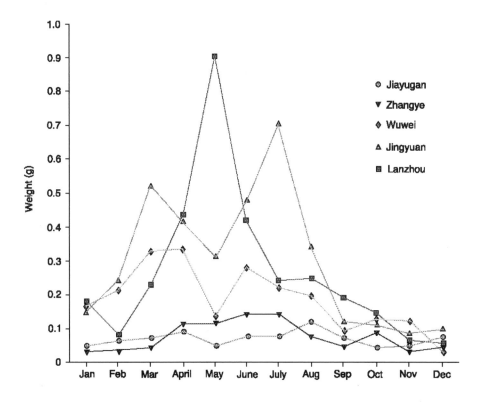

Figure 2.2: Mean monthly rate of dust deposition (1988–1991) at five stations along the Hexi Corridor, western Gansu province, China (after Derbyshire et al. 1998). Weight values x 30 = g per m².

Dust-fall sampling along the Hexi Corridor shows that while it is traditionally believed that dust storm activity peaks in spring to early summer, the pattern is more complex (Figure 2.2). Moreover, there is increased complexity down-wind along the length of the Hexi Corridor and across the Loess Plateau, a function of proximity to the dust sources and progressive increase in dust deposition leeward of the sources. For example, dust deposition amounts are relatively small and the spring and summer peaks only weakly developed in Jiayuguan (ca. 20 km WNW of Jiuquan) and Zhangye in the narrowest part of the western Hexi Corridor. However, dust accumulation rates rise significantly where the Corridor begins to widen just west of Wuwei, with resultant divergent airflow reaching highest values outside the Corridor on the western Loess Plateau at Jingyuan and Lanzhou. The double peak (March–April and June–July) is a clear feature of the dust deposition regime at Wuwei and Jingyuan (Fig. 2.2). Both lie adjacent to a complex of sources including the Hexi Corridor and the Tengger Desert, as well as the ice- and snow-fed Shiyang River (Wuwei) and the highly seasonal Yellow River (Jingyuan), but a full explanation of the double accumulation maximum requires more work. The curve for Lanzhou is dominated by a single substantial but broad maximum extending from March to August but with a clear peak in early summer (May). The location of Lanzhou toward the right (south) margin of the fan-shaped zone of dust deposition to leeward (east) of the outlet of the Hexi Corridor (see Fig. 4 in Derbyshire et al. 1998), and the possibility of additional dust sources in the northeastern Tibetan Plateau (as suggested by equivalent rates of loessic silt accumulation in the Xining Basin, about 200 km WNW of Lanzhou) may form part of the explanation of such a distinctive pattern. The reasons for such variations in accumulation patterns need to be better understood before a sound platform can be provided for the work of health scientists. The research required includes improved and longer-term dust-fall sampling, improved knowledge of ground-surface conditions in the source areas, and a fuller characterization of the meteorological controls, especially the windflow dynamics.

Dust suspended in the lower atmosphere is overwhelmingly of silt size. Minerogenic dusts sampled along the 1200-km long Hexi Corridor consist of finer silts derived from re-entrainment of loess, reworked silts, and loessic soils. There is no marked difference between summer and winter dusts. Both contain a mixture of fine to medium silts and silt-size aggregates of very fine silts and clay-size particles. Aggregates rich in clay-size (<2 μm) angular quartz are common. The finer respirable dust particles associated with the incidence of silicosis (<5 μm: Pendergrass 1958, <3 μm: Policard and Collet 1952) do not occur as discrete suspended grains in North China, but as silt-sized aggregates 5–50 μm in diameter, a notable proportion being 5–15 μm. The exposed population in China's semi-arid northwest is estimated to number 24 million people, although exposure data are limited. The highest rates of dust accretion appear to coincide with the thickest loess, indicating a dynamic relationship. Pilot measurements by medical staff of the Lanzhou Military Hospital (1982–1984) found outdoor and indoor dust concentrations of 4.2 mg/m^3 and 20 mg/m^3, respectively, within which free silica reached 61%. These very high dust loads occurred during storms, with analysis of the deposits showing important contributions from coarser fractions. In 1980, annual means of total suspended particulates (TSP) in Lanzhou of 1.14 mg/m^3 were reported for outdoor sites. Indoor/outdoor comparisons in Lanzhou in 1988 yielded TSP values of 0.94 and 0.76, respectively. Inhalable particulate levels of 3.02 mg/m^3 were reported in a rural home in Gansu Province in which dung was used as a cooking fuel. In the Minghai and Lianghua communes of the Hexi Corridor, the "average dust concentration" during 3 days in April 1991 (windy season) ranged from 8.25 to 22.0 mg/m^3 (15–26% free silica), according to Xu et al. (1993). We have identified only two surveys of silicosis in NW China, and these appear to be incomplete. In the study by Xu et al. (1993), radiographs were taken of 395 residents in a desert area, with a reported prevalence of 7% silicosis, rising to 21% among subjects over 40 years old. Such prevalences are comparable to those found in surveys of silicosis in industrial settings of which many have been published over the years both in developed and, now increasingly, in developing countries (Wagner 1997).

Radiological examination of the chest X-rays of 9591 residents, undertaken in Gansu province, showed a silicosis prevalence of 1.03%, rising to 10% in subjects over 70 years old [Personal communication, Changqi Zhou (Beijing)]. In addition, there are case reports of silicosis and silico-tuberculosis in farmers in the same province [Personal communication, Zhang Shi Fan (Lanzhou)]. There is thus some evidence that silicosis may be a public health problem in north China, although degree and extent cannot be quantified in the absence of statistics on morbidity levels across this extensive region. It should not be forgotten that these respiratory disease risks add to the substantial risk of nonsilicotic respiratory disease, the prevalence of cigarette smoking in Chinese adult males (over 15 years old) having exceeded 60% in 1984, with the figure of 7% for women smokers set to rise (World Health Organization 1997).

Wind-blown loessic dust, including aggregates of the finest fractions, is also known to be a health hazard in the Ladakh region of NW India. Chest X-ray surveys here, supported by analysis of limited necroptic lung tissue samples, suggest an incidence, among a randomly selected group of adults from two villages in the age range 50–62, of up to 45%, including three cases of progressive massive fibrosis (Norboo et al. 1991, Saiyed et al. 1991). Bulk chemical analysis of lung tissue showed that over 54% of the inorganic dust proved to be elemental silica (by X-ray diffraction) indicating that quartz constituted 16–21% of the extracted inorganic dust). Only 20% of men and none of the women were cigarette smokers, and silicosis was clearly shown to be of higher incidence in the village with the higher incidence of dust storms, suggesting an important link between the environment and the incidence of silicosis. Women's daily work is such that they are more exposed than men to fine dusts, and this is reflected in the greater incidence of advanced fibrotic lung disease in women in this region.

Finally, measures designed to reduce exposure to fine atmospheric dust have yet to be considered in High Asia, a region with a population inured to the difficulties of living for millennia in a generally harsh environment. Certain measures, such as the wearing of facemasks of doubtful efficacy, have been adopted by some sections of the population in some of the larger Chinese cities, but these are largely a response to perceived human-induced conditions ('pollution') rather than to natural phenomena.

Acknowledgments:

This summary owes much to the research of Dr. Tony Fletcher (London School of Hygiene and Tropical Medicine, U.K.), and Dr. Keith Ball (Middlesex Hospital, U.K.). I am also grateful for the help of my Chinese colleagues, Dr. Xingmin Meng (University of London, U.K.), Professor Shi Fan Zhang (Lanzhou, China) and Professor Changqi Zhou (Beijing, China). Dr. Tony Fletcher is also thanked for suggestions that improved an earlier draft.

References

Burbank, D.W., Leland, J., Fielding, E., Anderson, R.S., Brozovic, N., Reid, M.R. and Duncan, C. 1996. Bedrock incision, rock uplift and threshold hillslopes in the northwestern Himalayas. Nature, v.379, p.505–510.

Derbyshire, E., T.A. Dijkstra and I.J. Smalley (eds.) 1995. Genesis and Properties of Collapsible Soils. NATO ASI Series C: Mathematical and Physical Sciences, Vol. 468, Kluwer, Dordrecht, 413p.

Derbyshire, E. and Meng, X.M. 2000. Loess as a geological material. Chapter 3 in: Derbyshire, E., Meng, X.M. and Dijkstra, T.A. (eds.) 2000, Landslides in the thick loess terrain of Northwest China. John Wiley & Sons, Chichester and New York, p.47–90.

Derbyshire, E., Meng, X.M. and Kemp, R.A. 1998. Provenance, transport and characteristics of modern aeolian dust in western Gansu Province, China, and interpretation of the Quaternary loess record. Journal of Arid Environments, v.39, p.497–516.

Norboo, T. Angchuk, P.T. and Yahya, M. 1991. Silicosis in a Himalayan village population: role of environmental dust. Thorax, v.46, p.341–343.

Pendergrass, E.P. 1958 Silicosis and a few other pneumoconiosis: observations on certain aspects of the problems with emphasis on the role of the radiologist. J. of Roentgenology, v.80, p.1–41.

Policard, A. and Collet, A. (1952) Deposition of silicosis dust in the lungs of the inhabitants of the Saharan Regions, Arch. Industrial Hygienne and Occupational Med., v.5, p.527–534.

Saiyed, H.N., Sharma, Y.K. and Sadhus, H.G. 1991. Non-occupational pneumoconiosis at high altitude villages in central Ladakh. British Journal of Industrial Medicine , v.48, p.825–829.

Wagner, G.R. 1997. Asbestosis and silicosis. Lancet, v.349 (9061), p.1311–15.

Wang, Y.Y. and Song, H.L. (Eds) 1983 Rock Desert, Gravel Desert, Sand Desert Loess. Xi'an , Shaanxi People's Art Publishing House, N.P. World Health Organization 1997. Tobacco or Health: a Global Status Report. WHO.

Xu, X.-Z., Cai, X.-G., Men, X.-S. et al. 1993. A study of siliceous pneumoconiosis in a desert area off Sunan County, Gansu Province, China. Biomedical and Environmental Sciences, v.6, p.21–222.

3

Human Sickness and Mortality Rates
in Relation to the Distant Eruption of Volcanic Gases:
Rural England and the 1783 Eruption of the Laki Fissure, Iceland

JOHN GRATTAN
Institute of Geography and Earth Sciences
University of Wales, Aberystwyth, United Kingdom

MICHAEL DURAND
Natural Hazards Research Centre, Department of Geological Sciences
University of Canterbury, Christchurch, New Zealand

DAVID GILBERTSON
School of Conservation Sciences
Bournemouth University, Dorset, United Kingdom

F. BRIAN PYATT
Department of Life Sciences
Nottingham Trent University, Clifton Lane, Nottingham, United Kingdom

Introduction

Chapter 3 explores an apparent relationship between human mortality in England and exposure to acid volatiles derived from the Laki fissure eruption of 1783. It has long been known that volcanic tephra and gases may be transported great distances (Thórarinsson 1981). Research into their impacts on human health and the environment has typically focused on populations and environments relatively close to the eruption (e.g. Oskarsson 1980, Rose 1977, Thórarinsson 1979). However, recent investigations of documentary sources such as diaries and newspapers have suggested that in particular meteorological situations, and where air masses are stable, profound health and environmental consequences may have occurred in the British Isles and elsewhere in Europe, at great distances from the volcanic source in Iceland (Brayshay and Grattan 1999, Dodgshon et al. 2000, Durand 2000, Durand and Grattan 1999, Grattan 1998 a and b, Grattan and Brayshay 1995, Grattan and Charman 1994, Grattan and Pyatt 1994, 1999, Grattan et al. 1998, Stothers 1996). This chapter presents and examines documentary evidence for human illness, which may have been induced by volcanogenic air pollution, and mortality in several widely dispersed villages in rural England in the late eighteenth century. Burial records for these settlements point to a singular peak in mortality in the summer of 1783, a period that is coincident with the peak concentration of volcanic gases from the Laki fissure in the European environment.

Eruption Dynamics of Laki in 1783

The Laki fissure eruption took place between June 1783 and February 1784. It produced large quantities of acid volatiles — approximately ~120 Mt SO_2, 6.8 Mt HCl, and 15.1 Mt HF plus H_2S and NH_3. Of the total compounds emitted, approximately 60% were emitted during the first few months of activity and the majority of these emissions were confined to the troposphere (Sparks et al. 1997, Thordarson et al. 1996, Thordarson and Self 1993). The eruption therefore generated the largest known air pollution event of the last two millennia (Stothers 1996) and, moreover, one that was entirely natural in origin. A series of stable high-pressure air masses were stationed over northwest Europe throughout the summer of 1783 (Kington 1988). This meteorological situation resulted in the concentration of eruptive gases, which were manifested as a persistent, foul-smelling dry fog and various forms of acid-damage and were observed across the greater part of Europe during the summer of 1783 (Grattan et al. 1998, Stothers 1996) and perhaps even further afield in such far-flung locations as the Altai Mountains in China, Baghdad, and the coast of North America (Demarée et al. 1998, Stothers 1996, Thórarinsson 1979). The phenomena, which accompanied the dry fog, included a variety of forms of respiratory illness and acid damage to crops, trees, and bodies of water (Durand and Grattan 1999), the veiling and reddening of the sun and moon, and an intense smell of sulphur (Grattan and Brayshay 1995, Grattan and Pyatt 1995, Grattan et al. 1998).

Mortality and Human Health Impacts in 1783

The descriptions of human ill health in 1783 are remarkably consistent and link the presence of a strange dry fog or a strong sulphurous stench, with headaches, eye irritation, decreased lung function, and asthma. Several authors also link sudden epidemics and deaths to the sulphurous fogs; fuller descriptions are found in Durand and Grattan (1999) and inevitably more accounts come to light as new archival sources are located and consulted. The following selections are typical of many. In France "it (the fog) tires the eyes; and in Sallon, those people who have a weak chest, have endured some disagreeable symptoms . . . many people in the town have had headaches and . . . have lost part of their appetite" (Anon 1784). Elsewhere in France, Dreux (1783) described severe impacts on respiratory

health, to which he attributed a large number of deaths: "The parish of Champseru has been afflicted by a pestilential sickness. Patients were afflicted by a sickness of the throat. Many ignorant doctors treated it by bleeding and applying emetics and after 18 days there were 40 dead. One believes that the fogs of May, June, July and August that offended the sun and turned it red as blood forecast this curse. May God preserve my parish."

In the Netherlands, Brugmans (1787) and Swinden (1786) recorded similar problems. "After the 24th, many people in the open air experienced an uncomfortable pressure, headaches and experienced a difficulty breathing exactly like that encountered when the air is full of burning sulphur . . . asthmatics suffered to an even greater degree" (Brugmans 1787). "Between the 18th of June and the 21st of July the atmosphere was absolutely covered by the fog, and between the 22nd and 28th the air was very calm with little clouds. . . . Those people with weak chests experienced a similar sensation to that experienced when exposed to burning sulphur" (Swinden 1786). Descriptions in Britain were equally dramatic: "Such multitudes are indisposed by fevers in this country, that farmers have with difficulty gathered in their harvest, the labourers having been almost every day carried out of the field incapable of work and many die" (Cowper in King & Ryskamp 1981).

The descriptions presented above suggest that the noxious or toxic compounds within the volcanic dry fog may have been responsible for severe respiratory dysfunction in some people. The symptoms described suggest that the concentrations of SO_2 occasionally reached as high as 3000 $\mu g/m^3$. Bronchitis may be worsened when concentrations of SO_2 exceed 80 $\mu g/m^3$, while asthma is worsened at concentrations in excess of 572 $\mu g/m^3$ (Beverland 1998, Wellburn 1994) and the volcanogenic dry fog of 1783 clearly passed these critical thresholds for human health on many occasions. The concept that severe anthropogenic air pollution may cause respiratory illness and/or the death of vulnerable sections of the population is familiar in modern western societies (Dockery et al. 1992, Ostro et al. 1991, 1993, 1995, Pope and Kanner 1993, Pope et al. 1995). There are no compelling reasons to propose that volcanogenic air pollution, of sufficient concentration, may not have a similar impact on human health to anthropogenic air pollution.

The mortality data presented at right are drawn from the parish registers of three English rural parishes chosen at random. They cover the period from AD 1770 to AD 1795. In each case the data point to an atypical, singular, and profound increase in deaths in the summer months of 1783 — a feature that bears no obvious relationship to any longer term background trends. These three parishes clearly do not provide a statistically significant sample, yet they serve as a first test of an hypothesis that the prolonged and concentrated presence of volcanic gases in the boundary layer of the atmosphere over rural England in 1783 AD may have triggered widespread bronchial distress and death.

These data are gathered from the records of burials maintained in each parish. Typically these record the name of the deceased and the date of burial. Sadly for our purposes, they do not record cause of death or the age of the individual. Data are presented as the total deaths that occurred between July and September for the period 1770–1795.

In 1783 Cavendish in Suffolk was a small rural settlement of approximately 1000 inhabitants who made their living from maintaining woodland, dairying, and pig keeping. Between July and September 1783, eighteen people died in Cavendish set against a seasonal average for the period 1770–1795 of four. Eight people died in August and ten in September. (Fig. 3.1).

Castle Donington in Leicestershire was also an agricultural settlement of approximately 2000 people, who mainly

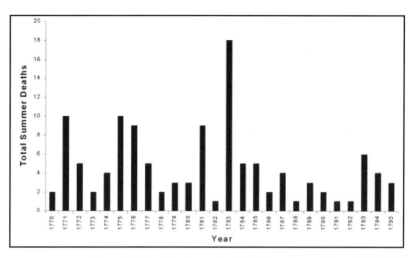

Figure 3.1: Total deaths occurring July–September — Cavendish, Suffolk

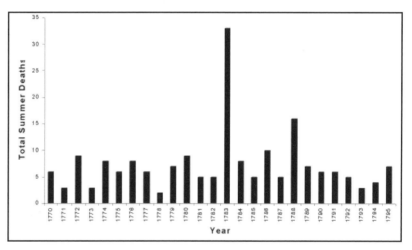

Figure 3.2: Total deaths occurring July–September — Castle Donington, Leicestershire

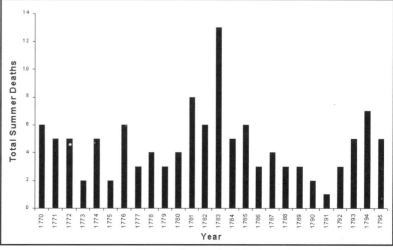

Figure 3.3: Total deaths occurring July–September — Banham, Norfolk

made their living from rearing and fattening livestock. Between July and September 1783, thirty-three people died in this parish, against an average for the period under review of seven. Eleven people died in August and sixteen in September. The deaths for the summer of 1783 are five standard deviations beyond the 1770–1795 average (Fig. 3.2).

In 1783, Banham in Norfolk was a small agricultural settlement of about 900 inhabitants making their living from pastures and pig keeping. Between July and September 1783, thirteen people died against an average of four (Fig. 3.3).

Castle Donington is approximately 150 km from Banham and Cavendish, which lie 40 km apart. While the numbers involved here may seem small to modern eyes, the summer of 1783 must have been a critical period for the parishes involved (see Fig. 3.4). These mortality patterns from three widely separated English rural parishes suggest that a crisis may have occurred during the summer of 1783 — awareness of physical process and the circumstantial evidence suggests acid volcanic gases may have been the key agent.

Discussion

Persistent and intense concentration of volcanic gases from Laki in the atmosphere of the British Isles in 1783 during the summer of 1783 has been demonstrated by considerable scholarship and is no longer in doubt. That these acid gases were present in sufficient concentrations to damage a wide range of tree and crop species and to cause respiratory health problems has also been suggested (Durand and Grattan 1999). Modern studies of anthropogenic air pollution incidents in and around major cities suggest that in addition to respiratory disorders similar to those described above, death rates may rise as vulnerable groups are affected by severe air pollution. In brief, the physiological stress caused by air pollution may reduce individual ability to cope with preexisting medical conditions, particularly cardiovascular problems. In the summer of 1783, the increased death rates and observations of respiratory disorders are co-incident in time with the eruption of the Laki fissure and the transport of volcanic gas and derived aerosols to Europe. The mortality figures presented here point to a potential link between the continental scale air pollution caused by the Laki

fissure eruption and a remarkable increase in death rates in central and eastern England during the summer of 1783. If not caused by air pollution, the mortality patterns presented still suggest an external influence, perhaps an epidemic of some nature, as yet unknown. However, all the contemporary accounts of illness reported in the summer months of 1783 point to air pollution as the key factor, and it is not unreasonable to propose that the volcanic gases may have had a role to play in these crises. A study of 400 parishes is now under way to test this hypothesis using data gathered by the Local Population Studies Centre at the University of East Anglia. Complementary data sets for elsewhere in Europe are also being sought.

Irrespective of the extent to which this past air pollution event was associated with increased death rates, it remains clear that the volcanic gases erupted from Laki were transported great distances through the atmosphere and probably had a significant impact on human health. Nor are such volcanic fogs as described here for 1783 rare; Camuffo and Enzi (1995) have described numerous similar events recorded in the Italian documentary record between the fourteenth

Figure 3.4: English parishes displaying anomalously high summer mortality in 1783.

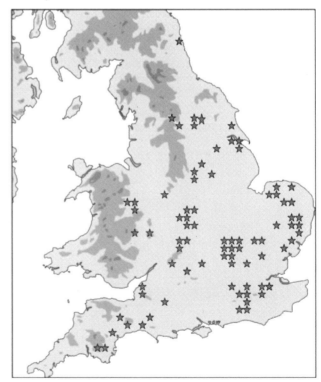

and nineteenth centuries, which are ascribed to eruptions of Italian volcanoes. The Laki fissure eruption and its impact on human health in Europe form a stark illustration of the need for scientists to explore the relationship between human health and all aspects of geological activity. In many areas of the world, air pollution is a serious problem; one need only consider the proximity of many rapidly growing cities to volcanic centers to conclude that volcanic gases will inevitably have a profound influence upon human health in the future (Durand and Grattan 2001).

References

Anon, 1784, Vue sur la Nature et L'origine du Bruillard. Journal du Physique, v.24, p. 8–17.

Beverland, I.J., 1998, Urban Air Pollution and Health. In. J. Rose (ed.) Environmental Toxicology: Current Developments p. 189–209. London, Gordon and Breach.

Brayshay, M. and Grattan. J.P., 1999, Environmental and social responses in Europe to the 1783 eruption of the Laki Fissure volcano in Iceland: A consideration of contemporary documentary evidence: Geological Society of London Special Publication, v.161, p. 173–188.

Brugmans, S. J. 1787. Naturkundige verhandeling over een zwavelatigen, nevel den 24 Juni 1783 inde provincie van stad en lande en naburige landen waargenomen. Leyden, Isaac van Campen, 51p.

Camuffo, D. and Enzi, S., 1995, Impact of clouds of volcanic aerosols in Italy during the last seven centuries: Natural Hazards, v.11(2), p. 135–161.

Demarée, G., Ogilvie, A. E. J. and Zhang, D., 1998, Further documentary evidence of northern hemispheric coverage of the Great Dry Fog of 1783: Climatic Change v.39, p. 727–730.

Dockery, D., Schwartz, J. and Spengler, J.D., 1992, Air pollution and daily mortality: Associations with particulates and acid aerosols: Environmental Research, v.59, p. 362–373.

Dodgshon, R., Grattan, J.P. and Gilbertson, D.D., 2000, Endemic stress, farming communities and the influence of volcanic eruptions in the Scottish Highlands: The Geological Society of London Special Publication, v.171, p. 267–280.

Dreux, M. 1783. cf. Rabartin, R. and Rocher, P., 1993, Les Volcans et la Révolution Française Paris,. L'Association Volcanologique Européene, 54p.

Durand, M., 2000, Impact of Volcanic Activity Upon the Air Quality and Environment of Europe: The University of Wales, Aberystwyth, Unpublished PhD Thesis.

Durand, M. and Grattan, J.P., 1999, Extensive Respiratory Health impacts of Volcanogenic dry fog in 1783 inferred from European documentary sources: Environmental Geochemistry and Health v.21, p. 371–376.

Durand, M. and Grattan, J.P., 2001, Volcanoes Air Pollution and Health. The Lancet, v.357, p. 164.

Grattan, J.P., 1998a, The distal impact of volcanic gases and aerosols in Europe: a review of the 1783 Laki fissure eruption and environmental vulnerability in the late 20th century: Geological Society of London: Engineering Geology Special Papers, v.15, p. 97–103.

Grattan, J.P., 1998b, The Toxicology of volcanic gases, their historical impact and their potential role in contemporary European environments: in Rose, J., ed, Environmental Toxicology: Amsterdam, Gordon Breach, p. 109–120.

Grattan, J.P. and Brayshay, M.B., 1995, An Amazing and Portentous summer: Environmental and social responses in Britain to the 1783 eruption of an Iceland Volcano: The Geographical Journal, v.161(2), p. 125–134.

Grattan, J.P, Brayshay, M. and Sadler, J.P., 1998, Modelling the impacts of past volcanic gas emissions: Evidence of Europe–wide environmental impacts from gases emitted in Italian and Icelandic volcanoes in 1783: Quaternaire, v.9(1), p. 25–35.

Grattan, J.P. and Charman, D.J., 1994, Non–climatic factors and the environmental impact of volcanic volatiles: implications of the Laki Fissure eruption of AD 1783: The Holocene, v.4(1), p. 101-106.

Grattan, J.P. and Pyatt, F.B., 1994, Acid damage in Europe caused by the Laki Fissure eruption — an historical review: The Science of the Total Environment, v.151, p. 241–247.

Grattan, J.P. and Pyatt, F.B., 1999, Volcanic eruptions, dust veils, dry fogs and the European Palaeoenvironmental record: Global and Planetary Change, v.21, p. 173–179.

King, J. & Ryskamp, C., 1981, The letters and prose writings of William Cowper. v.2. Clarendon Press, Oxford, 652p.

Kington, J.A., 1988, The weather for the 1780s over Europe. Cambridge, Cambridge University Press, 166p.

Oskarsson, N., 1980, The interaction between volcanic gases and Tephra: Fluorine adhering to tephra of the 1970 Hekla eruption: Journal of Volcanology and Geothermal Research, v.8, p. 251–266.

Ostro, B.D., Lipsett, M.J., Mann, J.K., Krupnick, A. and Harrington, W., 1993, Air pollution and respiratory morbidity among adults in Southern California: American Journal of Epidemiology, v.137, p.691–700.

Ostro, B. D., Lipsett, M.J., Wiener, M.B. and Selner, J.C., 1991, Asthmatic response to air-borne acid aerosols: American Journal of Public Health, v.81, p. 694–702.

Pope, C.A., Dockery, D.W. and Schwartz, J., 1995, Review of the Epidemiological Evidence of Health Effects of Air Pollution: Annual Review of Public Health, v.15, p. 107–132.

Pope, C.A. and Kanner, R.E., 1993, Acute effects of PM-10 pollution on pulmonary function of smokers with mild to moderate chronic obstructive pulmonary disease: American Review of Respiratory Disorders, v.147, p. 1336–1340.

Rose, W.I., 1977, Scavenging of volcanic aerosol by ash: atmospheric and volcanological implications: Geology, v.5, p. 621–624.

Stothers, R.B., 1996, The Great Dry Fog of 1783: Climatic Change v.32, p. 79–89.

Swinden, Mr., 1786, Observations sur quelques particularites meteorologiques del'année 1783. Memoires de l'academie Royale du Sciences, Années 1784–1785, p. 113–40.

Thórarinsson, S., 1979, On the damage caused by volcanic eruptions with special reference to tephra and gases: in Sheets, P.D. and Grayson, D.K., eds., Volcanic Activity and Human Ecology: New York, Academic Press, p. 125–159.

Thórarinsson, S., 1981, Greetings from Iceland: Ash falls and volcanic aerosols in Scandinavia: Geografiska Annaler, v.63A, p. 109–118.

Thordarson, Th. and Self, S., 1993, The Laki and Grímsvötn eruptions in 1783–85: Bulletin Volcanologique, v.55, p. 233–263.

Thordarson, Th., Self, S., Oskarsson, N. and Hulsebosch, T., 1996. Sulfur, chlorine and fluorine degassing and atmospheric loading by the 1783–1784 ad Laki eruption in Iceland: Bulletin of Volcanology, v.58, p.205–255.

Wellburn, A., 1994, Air Pollution and Climate Change: The biological impact 2nd Edition. Longman, Essex, 268p.

Part 1

Natural Geologic Hazards

B. Chemical — "Silent" and Long Term

The second group of chapters deals with what might be called "silent hazards" — those that are unseen and that may or may not be avoidable. These are hazards via inhalation or ingestion, the effects of which usually accumulate over time to cause a health problem when some threshold is exceeded.

One of the most direct yet hidden connections between health and geology is seen in the age-old practice of geophagy or pica, the eating of earth and clays for medicinal purposes. The inadvertent ingestion of soil, especially among children in many parts of the world, is quite well known. **Abrahams**[*] reminds us that this practice, common among certain animals, is also present in adults in many countries, such as Kenya and Uganda (see also **Davies**). The physiological implications of geophagy may have been important in the evolution of human diets (Johns and Duquette 1991), and in some cases there may be a nutritional benefit, such as making Fe and other elements available in deficient diets (Harvey et al. 2000). Recent studies have shown that geophagy in chimpanzees has dietary (e.g., mineral supplementation) and medical benefits (antacid and anti-diarrhea properties) and also works to detoxify dietary secondary metabolites such as alkaloids (Mahaney et al. 1999). Unfortunately, dirt also includes unhealthy organisms and other hazardous elements. Illustrating both sides of the interaction between human health and the natural background, geophagy, as Abrahams argues, deserves more attention by researchers seeking a greater understanding of the nutritional and pharmaceutical value of common soils, as well as their potential for harm when ingested. An example is the recent study by Smith et al. (2000) showing that Cu and Mg are less bioaccessible from geophagia of termite-nest and herbal remedy soils in Uganda than from ordinary soils ingested with the intake of regular foods and waters.

The chapters that follow deal with the many geogenic elements that are the normal constituents of the natural geochemical background, and which can occur in concentrations high (or low) enough to harm health. Of key importance in understanding elemental toxicity is bioavailability, that is, the restriction or magnification of element concentration during the transfer from source rocks, minerals, and soils to life forms. The many biochemical pathways, cycles, and mechanisms that control the transfer of elements and molecules up the food chain to plants and animals, and ultimately to humans, affect bioavailability. There

[*] *References to chapters in this volume are in **bold**.*

are a variety of "gates" along this chain where one element may compete with another for preferential uptake. Strontium (Sr), for example, is less likely to partition into bone mineral when the diet is replete with calcium (Ca) (Finkelman et al. 2001). Of course, some elements can be hazardous in even minuscule amounts, when the chemical form is both appropriate for transport from the geological to the human environment, and able to be absorbed by the body.

We start with the well-known poison arsenic (As), an element with a huge literature (over 3800 citations in GEOREF for 1997–2001). These publications encompass the geological context, the engineering aspects of water delivery, as well as the implications for policy decisions by governments coping with massive exposure. The current arsenic crisis in Bangladesh and other parts of Asia illustrates clearly the crossover between geology and health, and the following three chapters detail some of the very close associations.

Arsenic

Naidu and Nadebaum review the basic chemical principles involved in the dissolution and deposition of secondary As-bearing minerals in the environment following the weathering of primary minerals. Arsenic follows the normal and expected pathways via groundwater and surface water, to soils and crops, which cause debilitation, disease, and death for many people in the Bengal delta region. The authors outline and discuss some of the treatment methods proposed, since much of the domestic water is now coming from As-rich aquifers. Here is an obvious site for integrated investigations. A combination of geological mapping, and analysis of sediments and water from different localities is needed to establish baselines for natural levels of As in the environment. Cooperation with local agronomists to detect As in agricultural products, and with medical researchers, especially pathologists and dermatologists, would permit the identification of early stages of disease in those areas with high As concentrations. Tracking the ingestion of As-contaminated food and water requires organization and sharing of medical, agricultural, hydrological, and geochemical information, and analytical facilities to ensure that accurate and reproducible data are available. But remediation of the various geogenic sources of As cannot be accomplished without geological knowledge and information.

Finkelman et al. describe unexpected pathways for arsenic and fluorine (F) from domestically burned coal briquettes in Guizhou Province of China. The coal contains pyrite with trace amounts of As and other As-containing minerals. In addition to providing a general source of heat, coal burning is also used to dry various agricultural products in poorly ventilated circumstances. Several thousand people have been and are affected by arsenicosis, resulting both from the inhalation of As in the smoke and through ingestion, as chili peppers, a favorite constituent of the local diet, are dried over open fires and the plant is a bioaccumulator, concentrating the As. It is estimated that more than 10 million people in this area also suffer from some degree of fluorosis, due at least in part to F released during briquette burning. For these exposures it will be necessary to modify the traditional cultural practices of heating and drying systems of the local population, protecting them from the emissions of As and F.

Wang et al. continue this theme in their review of human diseases attributable to high exposure to As from air, diet, or drinking water. Observations show that many areas where there is chronic As exposure are low in selenium (Se). In a controlled experiment, analyses of water, blood, hair, and urine samples from three groups of patients in Inner Mongolia show that dietary Se appears to relieve symptoms, and there is a reduced concentration of As in the tissues. The medical data collected provides useful information to clinicians elsewhere who are dealing with As overload in their patients. As **Liṇdh et al.** point out, Se can also protect against mercury (Hg) toxicity. One point must be emphasized from the Wang et al. presentation. The analyses illustrate that inhaled or ingested As is distributed throughout the human body. The amount of As in hair, a "renewable resource," reflects the level of exposure and presents a sampling system over time that can be monitored relatively easily. Analyzing hair to determine As exposure has many advantages, as the famous case of the apparent poisoning of Napoleon demonstrated.

Though the results described here by Wang and his colleagues indicate a possible method of treatment, additional studies are needed to confirm the hypothesis that Se may be effective in preventing As bioaccumulations. One possible mechanism is that Se may prevent the substitution of As for sulphur (S) in the formation of certain proteins, such as the keratins. Laboratory testing should be expanded to humans as soon as practical if the hypothesis is upheld. Once the efficacy of such a preventive course of action is determined, medical teams should work in cooperation with geologists to identify natural sources of both As (hazard) and Se (remedy).

There is now no doubt about the direct connection between geogenic As, its transfer from arsenic-containing minerals in the geological environment via agricultural products, drinking water, or smoke, and the resultant debilitation and disease in millions of people. This geogenic hazard has only recently been recognized. At the time of this writing there is public discussion of a possible class action suit on behalf of the people in Bangladesh against the British Geological Survey (BGS). The claim is that during its earlier work in Bangladesh the BGS did not analyze waters for As and thus failed to warn of the health hazard. The BGS response so far is that only in the past 5 years or so has the need to include As in routine geochemical surveys of water quality been widely accepted — largely as a result of the Bangladesh experience (see <www.bgs.ac.uk/scripts/news/latest>). This situation emphasizes the need for geological organizations to be alert to the medical consequences of their work.

The three chapters on As suggest techniques for immediate amelioration of the As hazard, though the costs of even partial solutions may be prohibitive. Providing a safe source for domestic water is only one aspect of the issue in Bangladesh. Irrigation water is an equally important and potentially hazardous source over the long term via food ingestion. For the Chinese, ingesting As-laden chilies or continual inhalation of arsenical smoke will require major changes in personal habitat and habits. In addition, there may be serious problems in isolating the As-rich waste produced when As is removed from water supplies, for example, by sulphate-reducing bacteria (Fyfe et al. 2000).

Iodine

Iodine (I) as iodide (I^{1-}) is an essential element required for normal metabolism in humans. It is associated with the cascade of molecular species produced in the thyroid gland that governs growth and development. It has been long known that persons suffering I deficiency may appear physically normal but be mentally retarded. Ancient Chinese screens show individuals with enlarged necks (goiter) and a somewhat "crazy" appearance. Today, salt or cooking oil fortified with I provide the desirable daily intake of 100–300 µg/day of I, but as recently as 1996, the World Health Organization estimated that 29% of the world population was at risk for physical and mental defects, specifically retardation and cretinism, because of I deficiencies. In 129 countries it remains a significant health problem (USGS 1998).

Iodine is not in short supply, being a natural constituent of brines and nitrate deposits from which it is readily extracted. The ultimate source of iodine is sea water (approx. 0.05 ppm), and its transport in aerosol form to land is described for Norway by **Steinnes**, who shows that the concentration of I and Se in soils in coastal Norway is higher than in districts inland, both elements being concentrated on route via some unknown biological mechanism. He points out that in areas far from the sea, especially in developing countries, there may be unrecognized health problems related to deficiencies of I and Se. However, the conventional wisdom as to the source comes into question when **Fordyce et al.** discuss the complex mechanisms of iodine volatilization and transport in the island nation of Sri Lanka. Here, endemic goiter affects some 10 million people, but the relation between I, distance from the sea, and methods of plant uptake are still not clear. Rice, the great food staple of South Asia, does not appear to provide a significant pathway here.

Another important aspect to iodine in the environment concerns the long-lived radioactive isotope ^{129}I, which is created in nuclear accidents. Uptake of radioactive I by the human body is a potential cause of cancer and deserves careful study. Note that KI manufactured from naturally occurring I is prophylactic for anyone exposed to the radioactive species.

Mercury

Quicksilver (Hg) has been in use for hundreds of years and, although Hg vapor and all its compounds have toxic effects on vertebrate neurological systems, it is well to remember that Hg is also widespread in all that we eat (Boffetta et al. 1993). Only with concentration up the food chain does it become a serious health hazard. Minamata disease, which brought to world attention the dangers of industrial Hg pollution, is an example of a "silent" hazard, where the ingestion of a hazardous substance is undetected until disease and death result, and an investigation is mounted to identify the cause.

Mercury readily forms stable amalgams with most other metals, a fact that makes the element useful to miners extracting gold from placer deposits. Much of this work is carried out in small-scale operations where the Hg is carelessly handled, contaminating the local environment and causing a serious health hazard to miners and local populations alike. Mercury can enter rivers in the sediment load transported from placer operations to pose a threat to aquatic ecosystems and people living downstream. However, one study of such a situation in Brazil showed that the concentration of Hg in the surrounding sediments and rocks is natural, and they provide the major local source of the element (Telmer et al. 2000 and see **Berger**).

Another amalgam with mercury is of much more widespread concern, one that shows that medical geology is not confined to the effects of the geological environment on whole populations. This involves the use of Hg-containing amalgam in dental fillings, which has once again surfaced as a topic for worldwide attention (Spencer 2000). Specific details of the transport, deposition, and reactions of Hg in the dental environment are covered by **Lindh et al.**, who discuss whether every such filling is a potential source of Hg poisoning. This chapter and the investigations presented by Lindvall et al. describe analytical techniques utilized in dental medicine, cell biology, and metal biology. These applications may not be familiar to the geological reader, but metal binding to inorganic or organic species is essential for transport and for the construction of global geochemical cycles. Geoscientists employ scanning electron microscopy to define the distribution of elements in minerals and rock sections, and infer sources and transport mechanisms. Dental researchers use similar methods to detect, quantify, and determine the distribution of metal ions bound inside a cell, though the site and sample preparations are different. **Lindvall et al.** show that Hg from corrosion and wear of amalgam displaces zinc (Zn), with harmful effects on a range of proteins that need zinc to function properly. After removing the dental fillings, zinc levels in a test group of patients returned to normal levels. These chapters illustrate parallel techniques by biomedical scientists to study the complex internal pathways that metals may take when introduced into the human body. Perhaps geologists can further their investigations of transport of specific elements by examining the natural bioaccumulators and transport systems (see **Selinus**).

Cadmium and Molybdenum

Turning to the effects of natural toxicity on vegetation and animals, **Gough et al.** demonstrate the accumulation of cadmium (Cd) in Alaskan soils and thus in browse vegetation, particularly willow leaves and twigs ingested by wild animals. They calculate an enrichment factor, relate it to the uptake by plants, and suggest on the basis of comparisons with studies in Colorado that harm to herbivore health from Cd intake may be more widespread than previously recognized. **Frank** reviews work in Sweden and the U.S. on the causes of a diabetic wasting disease affecting Swedish moose as a result of increases in molybdenum (Mo) and decreases in Cu and Cd. Experimental studies on goats indicates a series of complex internal pathways and reactions involving the gut, blood, tissues, hypothalamus, and other organs. He suggests that *diabetes mellitus* may be a major factor in the deaths of the millions of domestic ruminants each year from Cu deficiency and molybdenosis.

These two papers provide interesting insights on the sources and restrictions for elements related to health. One documents the accumulation of Cd, while the other shows that it is in deficit for two groups of animals. The geochemical concentrations and bioavailability of particular elements need to be explored to address the disease potential for animals dependent on foraging. In this connection, a recent study by Hale et al. (2001)

shows that Cd accumulation in plant roots and its transport to shoots is determined not by the concentration of Cd in the soil, but rather by internal physiological processes. The levels of metals in soils and foods may thus give poor estimates of the dosage that finally reaches the plant cells or organisms that eat the plants. These considerations also apply to human populations constrained to geographic areas. There is a distinct possibility that the composition of food and water for some endemic disease situations will become a major consideration in identifying health hazards (see **Ceruti et al., Tatu et al.**).

References

Boffetta, P., Merler, E. and Vainio, H. (1993) Carcinogenicity of Hg and Hg compounds. Scandinavian Journal of Work, Environment and Health v.19, p.1–7.

Finkelman, R.B, Skinner, H.C.W., Plumlee, G.S., and Bunnell, J.E. 2001 Medical geology. Geotimes v.46, p.20–23.

Fyfe, W., J. Han and S. Koval, 2000. Sulphide minerals and bacteria clean arsenic polluted water: but where do we put the waste? Cogeoenvironment Newsletter v.17, p.15–16.

Hale, B., Berkelaar E., Chan, D., Black, W. and Johnson, D. 2001. Plants as modifiers of cadmium bioavailability. Geoscience Canada; v.28 (2), p.55–58.

Harvey, P.W., Dexter, P.B. and Darnton–Hill, I. 2000. The impact of consuming iron from non–food sources on iron status in developing countries. Public Health Nutr. v.3(4), p.375–383.

Johns, T and M. Duquette, 1991. Detoxification and mineral supplementation as functions of geophagy. Am. J. Clin. Nutr. v.53(2), p.448–456.

Mahaney, W.C., J. Zippin, M.W. Milner, K. Sanmugadas, R.G.V. Hancock, S. Aufreiter, S. Campbell, M.A. Huffman, M. Wink, D.Malloch & V. Kalm 1999. Chemistry, mineralogy and microbiology of termite mound soil eaten by the chimpanzees of the Mahale Mountains, Western Tanzania. Journal of Tropical Ecology, v.15, p.565–588.

Smith, B., B.G. Rawlins, M.J.A.R. Cordeiro, M.G. Hutchins, J.V. Tiberin, L Serunjogi, and A.M. Tomkins 2000. The bioaccessibility of essential and potentially toxic trace elements in tropical soils from Mukono District, Uganda. Journal of the Geological Society, v.157(4), p.885–891.

Spencer, A.J. (2000) Dental amalgam and mercury in dentistry. Australian Dental Journal v.45(4), p.224–234.

Telmer, K, M. Costa, E.S. Araujo, R.S. Angelica and Y. Maurice, 2000. Halting and repairing contamination – the importance of understanding source: alluvial gold mining and mercury pollution in the Tapajos River basin, Para, Brazilian Amazon, a case study. Cogeoenvironment Newsletter 17,, p.8–10.

USGS Minerals Yearbook 1998 Vol.1:38.1–38.8.

4

Human Geophagy:
A Review of Its Distribution, Causes, and Implications

Peter W. Abrahams
Institute of Geography and Earth Sciences
University of Wales, Aberystwyth, United Kingdom

Introduction

Geophagy (or geophagia) can be defined as the habit of eating clay or earth, a practice about which there is a great deal of misunderstanding. Though many people know and accept that geophagy is undertaken by wild and domesticated animals, and that humans can inadvertently ingest soil by (for example) hand-to-mouth activity, the deliberate consumption of soil by humans appears to be more difficult to comprehend. Yet geophagy (or pica) is a widespread among contemporary nonhuman primates (Krishnamani and Mahaney 2000), suggesting that the practice predates human evolution, and that soil ingestion has continued for a multiplicity of reasons. Therefore, although words such as 'evil,' 'odd,' 'filthy,' and 'degrading' have been applied to geophagy as practiced by humans (see Fig. 4.1), a more enlightened appraisal is to suggest that soil consumption should be considered to be

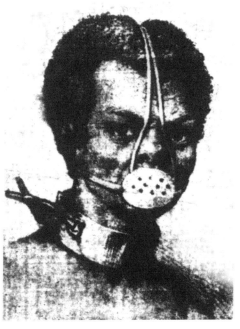

Figure 4.1: A rarely seen illustration of a slave forced to wear a face mask to prevent the eating of soil (Filho, 1946)

within the normal range of human behavior (Vermeer 1986).

Distribution of Geophagy

The oldest evidence for human geophagy comes from a prehistoric site at Kalambo Falls (Fig. 4.2) where the bones of Homo habilis, the immediate predecessor of Homo sapiens, have been found alongside a white clay believed to have been used for geophagical purposes (Root-Bernstein and Root-Bernstein 2000). Human migration then transferred geophagy to other parts of the earth, although Laufer (1930) concluded that the practice is not universal, being unknown in some countries such as Japan, Korea, and parts of Africa. To a certain extent this may be attributable to a lack of reporting on geophagy. For example, although the practice is not recorded in Namibia by Lagercrantz (1958, Figure 1), Thomson (1997) reports that the deliberate consumption of soil is commonly undertaken by preg-

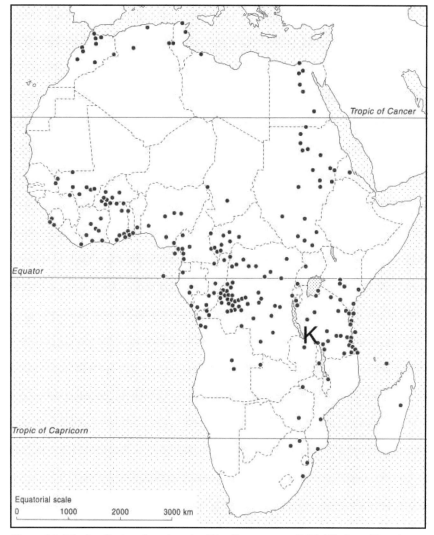

Figure 4.2: The distribution of geophagy in Africa (Lagercrantz, 1958). The letter K on the map marks Kalambo Falls where the oldest evidence of human geophagy has been discovered.

sume soil for a variety of purposes (Laufer 1930). Slaves shipped across the Atlantic were responsible for a large-scale transfer of the practice from Africa. Hunter (1973) states that soils were ingested by slaves for medical (including nutritional) and cultural reasons, but additionally large numbers of Negroes indulged in excessive soil-eating not only to become ill and to avoid work, but also to commit suicide in the belief that their spirit would return to their African homeland. So effective was this means of suicide (e. g., there are reports of some plantations being abandoned because of this practice) that slave owners were forced to employ harsh measures in an attempt to deter the practice (Fig. 4.1).

It was generally believed that the geophagy would cease with the termination of the slave trade, but the practice still persists among contemporary Afro-Americans who now eat soil for purposes other than as a means to commit suicide. Consequently, a number of reports have in recent decades commented on the geophagy undertaken by Afro-Americans in the rural southern areas of the United States. Here the practice is mainly to be found associated with Afro-American women and children of both sexes (Vermeer and Frate 1975), although Frate (1984) noted a relatively recent decline in geophagy among women of the region attributable to urbanization and the forces of modernization.

Europe has a long history of geophagy practiced by humans. For example, Thompson (1913) writes about Terra sigillata or Lemnian earth, a medicament used from about 100 BC until the middle of the nineteenth century, that had a great reputation throughout Europe.

nant women in the eastern Caprivi region of the country.

Geophagy can still be found relatively easily throughout many societies of the world today. While investigating geohelminth (i.e., parasitic worm) infection and mineral nutrient deficiencies in children and pregnant women, Geissler et al. (1997, 1998) indicated the prevalence of geophagy in parts of Kenya. In these studies, 73% of 285 school children aged 5–18 years indulged in the practice, whereas 154 of 275 pregnant women (56%) reported eating soil regularly. In the New World, geophagy was widespread previous to its discovery by Columbus, as indicated by early explorers and the numerous American tribes reported to con-

Figure 4.3: An example of soils that are consumed by humans today. These sikor tablets are ca. 3 cm square.

Today, while geophagy is not as common in Europe as certain locations elsewhere, the deliberate consumption of soil for medicinal or other reasons still continues, and indeed can be easily found by those wishing to investigate the practice (Fig. 4.3). For example heilerde (healing soil) can be purchased in Germany and used internally or externally. The cylindrically shaped objects in Figure 4.3 are examples of soil-based herbal remedies purchased in the central market of Kampala, Uganda (Abrahams,1997). Square tablets of so-called sikor (Fig. 4.3) are imported into the U.K. from Bengal and are consumed by children and pregnant women of the Asian community.

Causes and Implications

Although geophagy is not restricted to any particular age group, race, gender, or geographical region, the practice nevertheless is most common among certain groups of people such as childbearing women, children, and those from a poor socioeconomic background in developing countries (Abrahams and Parsons 1996). Often the soils consumed originate from specific locations (e.g., termite mounds in the tropics), and have certain qualities (e.g., color, flavor, and plasticity) that make them desirable to the consumer. The geophagical materials can be redistributed considerable distances from their original source of extraction. In a study of the *tierra santa* (sacred earth) clay tablets that are produced within Guatemala, Hunter et al. (1989) traced distribution channels throughout the country and into neighboring Belize, Honduras, and El Salvador. Some rudimentary manufacturing of the soil product may be evident; in this respect the sacred earths are ground, sieved, mixed with water, and pressed into a mold (to create a holy image) prior to drying and shipment to market.

The amount of soil ingested has been actually measured. Geissler et al. (1997) reports a median and range of soil intake of 28 g d^{-1} and 8–108 g d^{-1}, respectively by 5–18-year-old Kenyan school children. Abbey and Lombard (1973) estimated that 'a shoe box full a week,' a considerable quantity of soil, can be ingested. Sometimes eomania is encountered, whereby people develop a craving and uncontrollable urge for eating soil (Halsted 1968). The reason why soils are deliberately consumed can be difficult to establish, but known causative explanations include:

• The consumption of soil during times of famine. This has been recorded by numerous authors throughout the world (e.g., Loveland et al. 1989), with Laufer (1930) commenting that the pangs of hunger will be relieved since ingested soil will give a sensation of fullness to the stomach.

• The use of soil for detoxifying food. For example, Johns (1986) records the important detoxification function of geophagy in the domestication of the potato.

• The consumption of soil as a pharmaceutical. Soil may be the world's oldest medicine and is still used in treating gastrointestinal upsets, as an antidote to poisons such as herbicides, and as a traditional remedy, especially in developing countries (Fig. 4.3).

• The ingestion of soils for neuropsychiatric and psychological (comforting) reasons. Danford et al. (1982) report on geophagy practiced among members of an institutionalized mentally retarded population, while Layman et al. (1963) comment that women use geophagy for oral satisfaction in handling anxiety.

Ironically, many people assume that soils are ingested to satisfy a nutritional deficiency. Such a physiological explanation for geophagy is based on speculation and remains to be confirmed. Nevertheless soils do have the potential to supply mineral nutrients to the geophagist. For example, clays encountering the acidity of the stomach will release elements that they hold by cation exchange, for example, iron oxides will be partially soluble (Hendricks, cited by Cooper 1957). The ingestion of soil in particular can supply Fe in amounts that can account for a major proportion of

the recommended daily intake of this mineral nutrient (Abrahams 1997; Smith et al. 2000).

However, while soils can potentially provide certain benefits such as mineral nutrient supplementation to the consumer, some medical problems are likely if the geophagy is undertaken inappropriately through ingesting unsuitable or excessive quantities of soil. Such deleterious effects include:

• Mineral nutrient deficiency (Zn, potassium (K) and Fe deficiency have all been associated with geophagy).

• Toxicity as when soil has contaminants such as dioxin and lead (Pb).

• The ingestion of K–rich soil which has been associated with life–threatening hyperkalemia.

• Geohelminths swallowed with the soil causing an infection.

• Intestinal blockage and excessive tooth wear.

On occasion, deaths are reported. For example, a case of maternal death, resulting from a complete obstruction of the colon with subsequent perforation and sepsis, following geophagy is presented by Key et al. (1982), who comment that when the practice is unrecognized and untreated it can lead to serious complications.

Conclusions

Geophagy warrants more considered attention, although there is evidence that the practice is attracting renewed interest within the academic community. For the future an even greater realization of geophagy is required. Furthermore, there is a real need for multidisciplinary research involving a collaboration between earth scientists such as environmental geochemists and soil scientists, and those working in the medical profession such as specialists in mineral nutrition and those engaged in public health, pediatrics, and obstetrics. This is important since geophagy remains more widespread than most people realize, and ingested soils have important medical implications of which some such as the nutritional remain to be resolved and quantified.

References

Abbey, L. M., and Lombard, J. A., 1973, The etiological factors and clinical implications of pica: report of case: Journal of the American Dental Association, v. 87, p. 885–887.

Abrahams, P. W., 1997, Geophagy (soil consumption) and iron supplementation in Uganda: Tropical Medicine and International Health, v. 2, p. 617–623.

Abrahams, P. W., and Parsons, J. A., 1996, Geophagy in the tropics: a literature review: The Geographical Journal, v. 162, p. 63–72.

Cooper, M., 1957, Pica: Illinois, Charles C. Thomas, 105 p.

Danford, D. E., Smith, J. C., and Huber, A. M., 1982, Pica and mineral status in the mentally retarded: The American Journal of Clinical Nutrition, v. 35, p. 958–967.

Filho, A. M. D. L. R., 1946, O Rio de Janeiro Imperial: Rio de Janeiro, 494 p.

Frate, D. A., 1984, Last of the earth eaters: The Sciences, v. 24, p. 34–38.

Geissler, P. W., Mwaniki, D. L., Thiong'o, F., and Friis, H., 1997, Geophagy among school children in Western Kenya: Tropical Medicine and International Health, v. 2, p. 624–630.

Geissler, P. W., Shulman, C. E., Prince, R. J., Mutemi, W., Mnazi, C., Friis, H., and Lowe, B., 1998, Geophagy, iron status and anaemia among pregnant women on the coast of Kenya: Transactions of the Royal Society of Tropical Medicine and Hygiene, v. 92, p. 549–553.

Halsted, J. A., 1968, Geophagia in man: its nature and nutritional effects: The American Journal of Clinical Nutrition, v. 21, p. 1384–1393.

Hunter, J. M., 1973, Geophagy in Africa and the United States: a culture–nutrition hypothesis: Geographical Review, v. 63, p. 171–195.

Hunter, J. M., Horst, O. H., and Thomas, R. N., 1989, Religious geophagy as a cottage industry: the holy clay tablet of Esquipulas, Guatemala: National Geographic Research, v. 5, p. 281–295.

Johns, T., 1986, The detoxification function of geophagy and the domestication of the potato: Journal of Chemical Ecology, v. 12, p. 635–646.

Key, T. C., Horger, E. O., and Miller, J. M., 1982, Geophagia as a cause of maternal death: Obstetrics and Gynecology, v. 60, p. 525–526.

Krishnamani, R., and Mahaney, W. C., 2000, Geophagy among primates: adaptive significance and ecological consequences: Animal Behaviour, v. 59, p. 899–915.

Lagercrantz, S., 1958, Geophagical customs in Africa and among the Negroes in America: in Anell, B., and Lagercrantz, S., eds., Geophagical Customs: Studia Ethnographica Upsaliensia, v. 17, p. 24–84.

Laufer, B., 1930, Geophagy: Field Museum of Natural History Publication 280, Anthropological Series, v. 18, p. 97–198.

Layman, E. M., Mullican, F. K., Lourie, R. S., and Takahashi, L. Y., 1963, Cultural influences and symptom choice: clay eating customs in relation to the etiology of pica: Psychological Record, v. 13, p. 249–257.

Loveland, C. J., Furst, T. H., and Lauritzen, G. C., 1989, Geophagia in human populations: Food and Foodways, v. 3, p. 333–356.

Root–Bernstein, R., and Root–Bernstein, M., 2000, Honey, mud, maggots, and other medical marvels: the science behind folk remedies and old wives' tales: London, Pan Books, 280 p.

Smith, B., Rawlins, B. G., Cordeiro, M. J. A. R., Hutchins, M. G., Tiberindwa, J. V., Sserunjogi, L., and Tomkins, A. M., 2000, The bioaccessibility of essential and potentially toxic trace elements in tropical soils from Mukono District, Uganda: Journal of the Geological Society, London, v. 157, p. 885–891.

Thompson, C. J. S., 1913, Terra sigillata, a famous medicament of ancient times: 17th International Medical Congress, section 23 (History of Medicine), p. 433–444.

Thomson, J., 1997, Anaemia in pregnant women in eastern Caprivi, Namibia: South African Medical Journal, v. 87, p. 1544–1547.

Vermeer, D. E., and Frate, D. A., 1975, Geophagy in a Mississippi County: Annals of the Association of American Geographers, v. 65, p. 414–424.

Vermeer, D. E., 1986, Geophagy in the American South: Shreveport Medical Society, v. 37, p. 38.

5

Geogenic Arsenic and Associated Toxicity Problems in the Groundwater–Soil–Plant–Animal–Human Continuum

R. Naidu and P. R. Nadebaum
CSIRO Land and Water
PMB No. 2, Glen Osmond, Adelaide, South Australia, Australia

Introduction

Arsenic (As), a group (VI) element, is ubiquitous in the earth's crust, and therefore all living entities could be exposed. Significant adverse impacts of geogenic As on environmental and human health have been recorded in Bangladesh (Chowdhury et al. 1999), West Bengal, India (Chakraborty and Saha 1997), and China (Zheng et al. 1996). Although reported as early as the 1930s from the ingestion of As-contaminated water, it was not until severe human poisoning was recorded in the Indian subcontinent that the need for intensive research was recognized. This chapter explores several pathways of inadvertent exposure including ingestion of food, dust, soil, and dermal absorption, especially in the Asian region where wet land farming is a major activity among the farming community. We examine the source, release, and transfer of As from the local rocks and minerals to plants and humans via water and soil. Particular references are made to the incidences of health effects recorded in the Indian subcontinent as well as to some examples from other parts of the world.

Geochemistry and Biochemistry of Arsenic

Arsenic, the 20th most abundant element in the earth's crust, has two primary oxidation states under earth surface conditions: $As^{3+}(As^{III})$ and $As^{5+}(As^{V})$. The average abundance in the earth's crust is 1–2 mg kg^{-1} (NAS 1977). In sandstones, the As content has been reported to vary between 0.6 and 120 mg kg^{-1}, while in shales and clay formations the concentrations might be as high as 490 mg kg^{-1} (NAS 1977). Arsenic is also associated with new secondary minerals in sediments as a result of weathering. The element occurs naturally, forming arsenide minerals with the metals copper, lead, silver, and gold, and commonly in combination with, or in solid solution in, the common sulfides such as pyrite (FeS_2) or as arsenopyrite (FeAsS) (Yan Chu 1994, Gaines et al. 1997). Arsenic also occurs in reduced form as orpiment As_2S_3 and realgar AsS, and oxidized as the mineral arsenolite (As_2O_3), or its polymorph, claudite (also As_2O_3). The precise mineral species formed depends on the physio–chemical conditions at the time of mineralization. When sulfide mineral deposits, or coal that contains sulfides are mined and smelted (Zheng et al. 1996), As can be an inadvertent and unwanted part of waste material, tailings, and waste water. In many cases this unwanted element in soils and sediments is a result of human activities that accelerated its release.

Arsenite (H_3AsO_3) is the predominant species of As^{III} in solution, and adsorbs strongly on clays, iron oxides, and other soil components at neutral pH. Conversely, arsenate (H_3AsO_4) is the common form of As^V, and shows strong adsorption at the lower pH values 4 to 7 (Smith et al. 1998). Poorly ordered aluminosilicate minerals, such as allophane and imogolite, commonly found in soils derived from volcanic ash and spodosols, have strong affinity for As due to their high surface area and high anion exchange capacity (Bhattacharya et al. 2001). Under acid pH conditions both As^{III} and As^V containing ions bind specifically with iron and manganese oxyhydroxide minerals. At pH <4, these minerals dissolve, releasing the As. Further, because the stability of iron and manganese oxyhydroxide minerals is dependent on redox potential, under reducing conditions there may also be enhanced release of As. Aqueous As concentrations are strongly linked to the formation of soluble complexes under reducing conditions (Bhattacharya et al. 1997). In addition, As^V can be readily desorbed on the increase in pH, or in the presence of competing anions like PO_4, MoO_4, SO_4, and dissolved organic acids (Smith et al. 1998).

In groundwater, inorganic As may exist both as As^V and As^{III}. Through the action of bacteria, biomethylation, organic matter is degraded and As^V becomes the more soluble As species. Anoxic conditions in aquifers may also enhance the formation of methylated As based on the local conditions and bacterial species present. Any methylated As species produced may also sorb onto Fe-oxides with an affinity in the order As^V > dimethylarsinic acid (DMAA) > As^{III} > methylarsnic acid (MMAA) (Smith et al. 1998), and affect the distribution and mobilization of As. In aqueous environments, prokaryotes and eukaryotes bind inorganic As through production of DMAA and MMAA (Kramer and Allen 1988).

Toxicity of Arsenic

The toxicity of As varies with the nature of chemical species. As^{III} is considerably more toxic than As^V (Donohue and Abernathy 1999), and organo-As is the least toxic species found in the environment. Styblo et al. (1999) comment in an article describing work with cell cultures that the classification of As as a carcinogen has been exclusively based on epidemiological studies on residents of Taiwan, or elsewhere, where people were exposed to inorganic forms of As. Some soluble forms of As are readily absorbed by the body, whereas the less soluble forms are poorly absorbed and largely excreted in the feces. Organic forms that may be formed *in vivo* are virtually completely absorbed following digestion. The organo–arsine compound arseno-betaine [$(CH_3)_3AsCH_2COOH$] (Burrow 1987) is the major source of dietary As for people consuming large amounts of seafood. This form of As is nontoxic and is excreted unchanged in urine (Crecelius 1977). This suggests that people in the Indian subcontinent, whose major dietary intake of protein is from fish and other aquatic food, but have safe water supplies, may not be at risk from As poisoning — i.e., that there is a lower threshold below which No–Adverse–Effect of As is observed. In comparison, continued ingestion of water with As at 500 μg l^{-1} can cause chronic As disease (Abernathy et al. 1996) as the poisoning in Bangladesh and West Bengal, India, attest.

Although the mode of action of As varies with the nature of As species, the most common effect is through the inactivation of enzyme systems with As^V first reduced to As^{III} *in vivo* before exerting its toxic effect (Ginsburgh and Lotspeich 1963). These investigators also claim that As^V can compete with phosphate in the phosphorylation process, producing an arsenate ester of ADP. Irrespective of the nature of the toxicity mechanisms, evidence from reported incidences of arsenicosis throughout the world indicates that As in general (As^{III}, As^V, or organo-As) can prove deleterious to animal and human health.

Chronic As poisoning has been intermittently reported in the literature (Lu 1990), with the most common incidence induced by groundwater ingestion in six districts of West Bengal, India, and in Bangladesh. The most commonly observed symptoms of As poisoning are conjunctivitis, melanosis, and hyperkeratosis (Das et al. 1996; **Finkelman et al.**). Studies by Chakraborty and Saha (1987) documented elevated concentrations of As in hair, urine, and skin tissue of people living in West Bengal where groundwater is the main source of potable water. More than 30,000,000 Indians are drinking water containing As with concentrations greater than the Indian drinking water guidance value of 0.05 mg l^{-1} (Chowdhury et al. 1997; total As range: 0.05–3.7 mg l^{-1}). Das et al. (1996) report that at least 175,000 people have skin lesions caused by As poisoning in this region. However, Badal

(1998) says the number is over 400,000 with clinical symptoms of As toxicity, and millions of Bangladeshis have been recorded with skin lesions (Chowdhury et al. 1999). Exposure of local people to As contaminated groundwater has also been reported in Argentina, Chile, China, Ghana, Hungary, India, Mexico, Taiwan, Thailand, the United Kingdom, and the United States (BGS/MMI 1999). In the developed countries, alternative sources of water are tapped to avoid the As toxicity problems recorded in Bangladesh and West Bengal.

Sources of Geogenic Arsenic

The occurrence, origin, and mobility of As in groundwater is primarily influenced by local geology, hydrogeology, and geochemistry of the sediments. As in groundwater has been commonly related to the dissolution of sulfide minerals (Chowdhury et al. 2001), and the dissolution or desorption from iron oxyhydroxides (Bhattacharya et al. 1997 , Nickson et al. 1998) or upflow of geothermal water (Welch et al. 1999).

Weathering of Primary Minerals

Arsenic is released in the soil environment following weathering of the common minerals arsenopyrite (FeAsS), and pyrite (FeS_2) containing As, or any other primary sulfide minerals. When pyrite is exposed to oxygen and water, the mineral weathers to form As^V,

Fe^{III} and SO_4. The overall reaction describing the oxidation of pyrite and arsenopyrite can be written as Equations 1 and 2 (Table 5.1).

The half redox reactions are written as Equations 3–6 (Table 5.1) in which S, Fe, and As are oxidized using O_2 as the electron acceptor. The weathering reactions are influenced by the: (1) presence of moisture, (2) pH, (3) composition of pore water, (4) temperature, (5) carbon dioxide equilibria, (6) solubility, and (7) redox values of the surrounding media. In summary, the oxidation of the most common mineral, As-rich pyrites, is the major source of As in groundwater in the vast tract of alluvial aquifers within the Bengal Delta Plain in Bangladesh and West Bengal (Bhattacharya et al. 2001). In addition to these reactions, biotic activity may contribute to pyrite dissolution. Following release of As from the As-containing minerals, the element can be transported by a range of physical, as well as chemical, processes (Garrels and Lerman 1977).

During the oxidation of pyrite, the released reduced Fe may be oxidized via both biotic and abiotic pathways. Abiotic Fe^{2+} oxidation is slow, but Fe^{2+} oxidation by microorganims is common in natural waters and sediments. Once oxidized to Fe^{3+} [Fe^{III}], the insoluble ferric hydroxides precipitate. The precipitation of ferric hydroxide can be expressed by the reaction in

---

Table 5.1: Weathering reactions

1. $4FeAsS_{(s)} + 14O_2 + 16H_2O \Leftrightarrow 4SO_4{}^{2-}{}_{(aq)} + 4AsO_4{}^{3-}{}_{(aq)} + 4Fe(OH)_3 + 20H^+{}_{(aq)}$

2. $2FeS_{2(s)} + 7H_2O + 7.5O_2 \Rightarrow 2Fe(OH)_{3(s)} + 8H^+ + 4SO_4{}^{2-}{}_{(aq)}$

3. $O_2 + 4 H^+{}_{(aq)} + 4e^- \Rightarrow 2H_2O$ $\quad\quad\quad\quad$ E° = 1.23V

4. $S^{2-}{}_{(aq)} + 4H_2O - 8e^- \Rightarrow SO_4{}^{2-}{}_{(aq)} + 8H^+{}_{(aq)}$ $\quad\quad$ −E° = −0.76V

5. $AsO_2{}^-{}_{(aq)} + 2H_2O - 2e^- \Rightarrow AsO_4{}^{3-}{}_{(aq)} + 4H^+{}_{(aq)}$ $\quad$ −E° = −0.56V

6. $Fe^{2+}{}_{(aq)} - e \Rightarrow Fe^{3+}{}_{(aq)}$ $\quad\quad\quad\quad\quad\quad$ E° = − 0.77 V

7. $Fe^{3+}{}_{(aq)} + 3H_2O_{(l)} \Leftrightarrow Fe(OH)_{3(s)} + 3H^+{}_{(aq)}$

Equation 7 (Table 5.1) which has critical importance for retention and mobilization of As in soils as well as in aqueous media. These exceedingly fine-grained gel–like ferric oxyhydroxides solids are an important phase in sorption of any released As.

Both As^V and As^{III} are adsorbed on $Fe(OH)_3$, with the affinity higher for As^V as compared to As^{III}, and much lower for the organo-As species. Similar to many other oxyanions, the amount of total As sorbed by ferric hydroxide varies significantly with pH. Maximum adsorption for As^{III} is at pH 7.0, while the As^V maximum is at pH 4.0 (Smith et al. 1998). When the pH < 4 the acidophilic pyrite oxidizing bacteria *Thiobacilus ferrooxidans* can increase the rate of pyrite oxidation by several orders of magnitude. The biogenic effect may be mediated by other species as well.

Although the precise nature of mechanisms contributing to As in groundwater in the Indian subcontinent and specifically in the sedimentary aquifers of the Bengal Delta Plain are still not clear, two hypotheses have been proposed. The release of As takes place (a) during the oxidation of As-rich pyrite (FeS_2) or arsenopyrite (FeAsS) due to lowering of the water table during groundwater pumping (Saha and Chakraborty 1995) or through (b) desorption from or reductive dissolution of Feoxyhydroxides in a reducing aquifer environment (Bhattacharya and Jacks 2000, Bhattacharya et al. 1997, Nickson et al. 1998, Nickson et al. 2000, USAID 1997). According to Chowdhury et al. (2001), the As in West Bengal is natural (geogenic) and associated with pyrites in the alluvial sediments alongside the river Ganges. Groundwater withdrawal for agricultural irrigation leads to the oxygenation of the aquifer, decomposes the pyrite (FeS_2) rich in As, and the acid formed aids the release of As in a soluble form to the groundwater (Table 5.1: Equations 1 and 2). This is consistent with the suggestions of Welch et al (1988) that "mobilization of As in sedimentary aquifers may be in part a result of changes in the geochemical environment due to withdrawal of water for agricultural irrigation."

Dissolution of Oxides

Secondary minerals or salts that have adsorbed or incorporated other ions can increase the concentration of As in soils and groundwater. Work in Bangladesh and West Bengal suggests that iron oxyhydroxides, secondary minerals, are the major phase associated with As (Bhattacharya et al. 1997, Nickson et al. 2000). If the aquifers formed under aerobic conditions and sorbed As as arsenate onto iron and aluminum oxyhydroxides, but later became anaerobic, the reducing conditions would lead to dissolution of the oxyhydroxides and any associated As would be released into the groundwater. Such a scenario is common in many parts of the world (Bhattacharya et al. 1997). In the subsurface environment, the mobility of As is especially enhanced under anoxic conditions, making the risk for contamination higher in groundwater as compared to the surface water sources such as rivers, lakes, and reservoirs. The methylation of inorganic As to MMAA and DMAA, and changes in the biogeochemical microenviroment within the aquifers, are also possible. However, there is limited evidence of the presence of such species in groundwater samples in the Indian subcontinent.

Irrigation as a Source of Geogenic As in Surface Soils

It is likely that the use of groundwater for irrigation has led to or will lead to extensive contamination of soils throughout West Bengal and Bangladesh, although there is very little information on this. Similar contamination occurs in Mongolia and China where groundwater drawdown has occurred with agricultural demand. Limited studies by CSIRO Land and Water in Australia show that concentrations of As in surface soil in areas of long term irrigation in Bangladesh can exceed the guidance value of 100 mg kg^{-1} for residential soils. We can calculate the As load associated with irrigation by groundwater. Preliminary studies from two villages and a limited set of tubewell samples suggests annual deposition rates exceeding 4 tons of As per hectare.

Pathways of Human Exposure to Geogenic As

The major pathway for geogenic As exposure via the use of groundwater for domestic (drinking, cooking) purposes has been reported from several areas of the world (Bhattacharya et al. 1997, Nriagu 1994). Ingestion of crops or animals raised on either contaminated soils or with contaminated groundwater, as well as dermal absorption of As during wetland rice farming, may also lead to significant exposure to As. Only very few data have been gathered to assess these pathways.

Based on studies in Denmark, U.K., and Germany, Weigert et al. (1984) have shown that in countries where geogenic As is not present at high levels, the average intake of As via food of plant origin is 10–20 μg As day^{-1}, which is equivalent to only 10–12% of the estimated total dietary intake of As in these 3 countries. This range is similar to the recent survey by Schoof et al. (1999), which found that, for most adults, dietary inorganic As intake is of the order of 1–20 μg day^{-1} However, such investigations do not present a realistic picture of As intake in countries like Bangladesh and West Bengal, where vegetables are produced on soils subjected to over 20 years of irrigation with As-contaminated groundwater. Further, in other areas not employing As-containing irrigation water, the soils used for cropping elsewhere may have been transported from the As-rich sites. Cycling of As from parent material to soils to waters for irrigation, or surficial transport of soils containing As, eg. via erosion or deliberate shifting to augment local productivity by the farmers, could significantly elevate the total As in these regions (Naidu and Skinner 1999). The significance of geogenic As ingested through the food route is not known. In addition, incidental ingestion and inhalation of dust or indoor air derived from combustion of As contaminated materials could also be a significant exposure pathway (**Finkelman**).

Options for Minimizing Ingestion of Geogenic As

The recent incidences of As poisoning in many Asian countries has led to increased research on technologies for remediation of groundwater or for alternate options for potable water. While many treatment techniques (e.g., ion exchange, reverse osmosis, coagulation) for removing As were developed prior to the poisoning reported in these countries, their poor economy and indequate infrastructure have limited their applications. Most of the treatment technologies are based on As^{V} reacting with either an oxidic mineral or ion exchange surface in which the initial step in the process transforms As^{III} to As^{V}, which is either adsorbed on reactive surfaces or precipitated (Montgomery 1985). Oxidants such as chlorine (Cl), permanganate, solar radiation, and oxygen (O) have been commonly used (see Table 5.2).

In contrast to treatment of As-contaminated water, little effort has been directed towards remediating As-contaminated soils. Current studies for managing As-contaminated soils include soil washing (Naidu et al. 1999), chemical fixation (van der Hoek and Comans 1996), bioremediation (Frankenberger and Losi 1995), and electro-kinetic remediation (Lageman 1993).

At present no technique is available for minimizing uptake of As by crops, and it will require a comprehensive approach to address all the various pathways delineated above. An important tool for minimizing the effect of geogenic As would be to develop a map that delineates the distribution (and therefore the potential risk) posed by geogenic As in groundwater, soils, and crops at the landscape level. In this way appropriate strategies for minimising exposure could be tailored to specific areas. One could imagine the provision of water treatment or alternative water sources where water is contaminated, or the crops selectively distributed to avoid high As environments by relocating, covering, or diluting the surface soil, or by using more advanced methods such as phytoremediation. Continual upgrading of such maps could track changes in bioavailability and concentration of As over time, since ageing can decrease the bioavailability of contaminants, and any soil disturbance such tilling will gradually reduce localized surficial contamination.

Conclusions

Exposure of people to geogenic As in Bangladesh represents one example of a major hazard related to human manipulation of the natural environment. The present focus is on controlling exposure to As by either treating the groundwater before human use, which is technically possible, or providing alternative clean sources of water. Strategies for treatment that are affordable, cost-effective, and socially acceptable to all of the affected people, need more research. Importantly, there has been little consideration of alternative exposure pathways (the food and soil contamination), which can be relatively easily assessed with present analytical techniques. The challenge for the future will be to quantify these exposure routes and to devise and provide appropriate management strategies.

Table 5.2: Some treatment technologies for removal of As from water

Treatment Technology	Mechanism	Reference	Remarks
Coagulation	As^{III} +oxidant $\Rightarrow As^V$ As^V +coagulant (ferric sulfate, alum, lime) $\Rightarrow$ immobilization of As	Montgomery(1985), Paige et al. (1996)	Commonly used to remove As from drinking water in many countries
Activated alumina	As^{III} + oxidant $\Rightarrow As^V$ As^V + activated alumina $\Rightarrow$ adsorption	Gupta and Chen (1978)	Technology available with potential for regeneration of the surface
Ion exchange	Strong-base anion exchange resin	Montgomery (1985)	Resin can be recycled — technology commonly used
Reverse osmosis	Membrane technology; uses external pressure to reverse natural osmotic flow; polyvalent oxy-anions such as As are excluded in this process	Montgomery (1985)	Can be an expensive process
Fe-oxides	Based on the adsorptive capacity of these minerals; uses Al_2O_3 and/or TiO_2 quoted with newly precipitated $Fe(OH)_3$	Hlavay and Polyák (1997)	Used by many water treatment plants
Biotic-abiotic treatment	Involves oxidation of As^{III} to As^V followed by treatment with Fe-salt	Ghosh and Yuan (1987); Korte and Fernando (1991)	Emerging technology
Aquifer oxygenation	Air or O_2 injection into the aquifer to oxidize As^{III} to As^V to enhance precipitation in the aquifer	Frisbie (1996)	Attractive process but could be expensive
Passive reactive barriers	Reactive barrier system made up of natural clay/oxidic system	Naidu et al. (2000)	Expensive to install but cheap in the long term
Oxidic soils			
Three pitcher system	Oxidation of Fe^{2+} to Fe^{3+} and subsequent precipitation of Fe–OOH and removal of As via adsorption on to the oxide surface	Chowdhury et al. (2001)	Cheap option

References

Abernathy, C.O., Chappell, W.R., Meek, M.E. (1996) Roundtable summary: Is ingested inorganic arsenic a 'threshold' carcinogen? Fund. Appl. Toxicol., v.29, p.168–175.

Badal, (1998) Status of arsenic problem in two blocks out of sixty in eight groundwater arsenic affected districts of West Bengal, India. PhD thesis, School of Environmental Sciences, Jadavpur University, Calcutta, India, p. 208.

BGS/MMI (1999). Groundwater Studies for Arsenic Contamination in Bangladesh, Volumes S1, S2, and Main Report, British Geological Survey, Nottingham, England.

Bhattacharya, P., Chatterjee, D., and Jacks G. (1997) Occurrence of arsenic contaminated groundwater in alluvial aquifers from Delta Plains, Eastern India: Options for safe drinking water supply. Int. Jour. Water Resources Management v.13(1), p.79–92.

Bhattacharya, P., and Jacks, G. (2000) Arsenic contamination in groundwater of the sedimentary aquifers in the Bengal Delta Plains: A review. In: P Bhattacharya and AH Welch, eds. Arsenic in Groundwater of Sedimentary Aquifers. Pre–Congress Workshop, 31st Int Geol Cong, Rio de Janeiro, Brazil, 2000, p.18–21.

Bhattacharya, P., Frisbie, S.H., Smith, E., Naidu, R., Jacks, G., and Sarkar, B. (2001) Arsenic in the Environment: A Global Perspective. In Handbook of Heavy Metals in the Environment. Sarkar, B. (ed). Marcel Decker Inc. New York. In press.

Burrow, M. (1987) Trace Element and Analytical Chemistry in Medicine and Biology. (Eds. Brätter, P and Schramel, P. Berlin; New York: W de Gruyter.

Chakraborty, A.K., and Saha, K.C. (1987) Arsenical dermatosis from tube well water in West Bengal. Ind. J. Med. Res. v.85, p.326–334.

Chappell, W.R., Abernathy, C.O., and Calderon, R.L. (1999) Arsenic Exposure and Health Effects: Proceedings of the Third International Conference on Arsenic Exposure and Health Effects, July 12–15, 1998, San Diego, California. Elsevier. p 416.

Chowdhury, T.R., Mandal, B.K.R., Samanta, G., Basu, G.Kr., Vhowdhury, P.P., Canada, C.R., Karan, N.Kr., Lodh, D., Dahr, R.Kr., Das, D., Saha, K.C. and Chakraborty, D. (1997) Arsenic in groundwater in six districts of West Bengal, India: the biggest arsenic calamity in the world: the status report up to August, 1995. (Eds. CO Abernathy, RL Calderon and WR Chappell) Chaman and Hall, p. 93–111.

Chowdhury, U.K., Biswas, B.K., Dhar, R.K., Samanta, G., Mandal, B.K., Chowdhury, T.R., Chakraborty, D., Kabir, S., and Roy, S. (1999) Groundwater arsenic contamination and suffering of people in Bangladesh. In: WR Chappell, CO Abernathy and RL Calderon (eds) Arsenic Exposure and Health Effects: proceedings of the 3rd International Conference on Arsenic Exposure and Health Effects, July 12–15, 1998, San Diego, California. p 165–182, Elsevier.

Chowdhury, U.K., Rahman, M.M., Paul, K., Biswas, B.K., Basu, G.K., Chanda, C.R., Saha, K.C., Lodh, D., Roy, S., Quamruzzaman and Chakraborty, D. (2001) Groundwater arsenic contamination in West Bengal, India, and Bangladesh: Case study on bioavailability of geogenic arsenic. In: R Naidu, VVSR Gupta, SR Rogers, N Bolan, and D Adriano (eds) Bioavailability, Toxicity and Risk Relationships in Ecosystems. IBH/Oxford, New Delhi, India (In Press).

Crecelius, E.A. (1977) Changes in the chemical speciation of arsenic following ingestion by man. Environ. Helth Perspect. v.19, p.147–150.

Das, D., Samantha, G., Mandal, M.K. (1996) Arsenic in groundwater in six districts of West Bengal, India. Environ. Geochem. and Health, v.18, p.5–15.

Donohue, J.M. and Abernathy, C.O. (1999) Exposure to inorganic arsenic from fish and shellfish. In: W.R. Chappell, C.O. Abernathy and R.L. Calderon (eds.) Arsenic Exposure and Health Effects: proceedings of the 3rd International Conference on Arsenic Exposure and Health Effects, July 12–15, 1998, San Diego, California. p 89–98, Elsevier.

Finkelman, R.B., Belkin, H.E., Zheng, B., and Centeno, J.A. (2000). Arsenic poisoning caused by residential coal combustion in Guizhou Province. In: P. Bhattacharya, A.H. Welch, eds. Arsenic in Groundwater of Sedimentary Aquifers. Pre–Congress Workshop, 31st Int Geol Cong, Rio de Janeiro, Brazil, p. 41–42.

Frankenberger, W.T., and Losi, M.E. (1995) In: H.D. Skipper, R.F. Turco, eds. Bioremediation. Madison, WI: Soil Sci Soc Am Special Publication 43, p. 173–210.

Frisbie, S.H. Aquifer Oxygenation for Arsenic Removal from Groundwater at a Superfund Site in New England. Confidential Report, 1996.

Gaines, R., Skinner, H.C.W., Foord, E., Mason, B., and Rosensweig, A. (1997) Dana's new mineralogy: the system of mineralogy of James Dwight Dana and Edward Salisbury Dana. 8th ed., Publisher: New York : Wiley.

Garrels, R.M., and Lerman, A. (1977) The exogenic cycle: reservoirs, fluxes and problems. In: Global Chemical Cycles and Their Alteration by Man, Dalhem Workshop. p.23–31.

Ghosh, M.M., and Yuan, J.R. (1987) Adsorption of inorganic arsenic and organoarsenicals on hydrous oxides. Environmental Progress :p.150–157.

Ginsburg, J.M., and Lotspeich, W.D. (1963) Interrelations of arsenate and phosphate transport in the dog kidney. Amer. J. Physiol., v.205, p.707–714.

Gupta, S.K., and Chen K.Y. (1978) Arsenic removal by adsorption. Journal Water Pollution Control Fed. v. 50, p. 493–506.

Hlavay, J., and Polyák. K. (1997) In: C.O. Abernathy, R.L. Calderon, W.R. Chappell, eds. Arsenic: Exposure and Health Effects. New York: Chapman & Hall. p. 383–405.

Korte, N.E. and Fernando, Q. (1991) A Review of Arsenic (III) in Groundwater Cri Rev Env Control v.21, p.1–39, 1991.

Kramer, J.R., and Allen, H.E. (1988) Metal Speciation: Theory, Analysis and Application. Chelsea, MI: Lewis Publishers.

Lageman, R. (1993) Electroreclamation: Applications in the Netherlands. v.27, p.2648–2650.

Montgomery, J.M. (1985) Water Treatment Principles and Design. New York, NY: John Wiley & Sons.

Naidu, R., and Skinner, H.C.W. (1999) Arsenic contamination of rural ground water supplies in Bangladesh and India: Implications for soil quality, animal and human health. In: C. Barber, B. Humphries and J. Dixon (eds) Proceedings International Conference on Diffuse Pollution, 16–20 May, 1999, Perth. p.407–417.

Naidu, R., Smith, E., Smith, J., and Kookana R.S. (2000) Is there potential for using strongly weathered oxidic soils as reactive landfill barriers? Abstracts "Towards Better Management of Wastes and Contaminated Site in the Australasia–Pacific Region Workshop" Adelaide, South Australia 3–5 May 2000.

Naidu, R., Smith, J., and Swift, R.S. (1999) Soil washing techniques for remediation of arsenic contaminated soils. In: Proceedings of 5th International Conference on the Biogeochemistry of Trace Elements, July 11–15 (1999) pp.1026–1027, Vienna (Eds. W.W. Wenzel, D.C. Adriano, B. Alloway, H.E. Doner, C. Keller, N.W. Lepp, M. Mench, R. Naidu, and G.M. Pierzynski).

NAS (National Academy of Sciences) (1977) Medical and Biologic Effects of Environmental Pollutants: Arsenic.

Nickson, R., McArthur, J., Burgess, W., Ahmed, K.M., Ravenscroft, P., Rahman, M. (1998) Arsenic poisoning of Bangladesh groundwater. Nature v.395, p.338–.

Nickson, R.T., Mcarthur, J.M., Ravenscroft, P., Burgess, W.G., Ahmed, K.M. (2000) Mechanism of arsenic release to groundwater, Bangladesh and West Bengal Applied Geochemistry v.15, p.403–413.

Nriagu, J.(ed.) (1994) Arsenic in the Environment, Part II: Human Health and Ecosystem Effects. John Wiley & Sons. 293p.

Paige, C.R., Snodgrass, W.J., Nicholson, R.V., and Scharer, J.M. (1978) The crystallization of arsenate–contaminated iron hydroxide solids at high pH. Wat Env Res v.68, p. 981–987, 1996.

Saha, A.K. and Chakraborty, C. (1995). Geological and geochemical background of the arsenic bearing groundwater occurrences of West Bengal. Proceedings, Int Conf on Arsenic in Groundwater, Calcutta, India.

Schoof, R.A., Eickhoff, J., Yost, L.J., Crecelius, E.A., Cragin, D.W., Meacher, D.M., and Menzel, D.B. (1999) Dietary exposure to inorganic arsenic. In: WR Chappell, C.O. Abernathy and R.L. Calderon (eds) Arsenic Exposure and Health Effects: proceedings of the 3rd International Conference on Arsenic Exposure and Health Effects, July 12–15, 1998, San Diego, California. p 81–88., Elsevier.

Smith, E, Naidu, R and Alston, AM (1998). Arsenic in the Soil Environment: A review. Adv Agron v.50, p.149–195.

Styblo, M, Vega, L, Germolec, D.R., Luster, M.I., Razo, L.M.D., Wang, C, Cullen, W.R. and Thomas, D.J. (1999) Metabolism and toxicity of arsenicals in cultured cells. In: W.R. Chappell, C.O. Abernathy and R.L. Calderon (eds.) Arsenic Exposure and Health Effects: proceedings of the Third International Conference on Arsenic Exposure and Health Effects, July 12–15, 1998, San Diego, California. p 311–323., Elsevier.

USAID (United States Agency for International Development). 1997. Report of the impact of the Bangladesh Rural Electrification Program on groundwater quality. Prepared by the Bangladesh Rural Electrification Board. USAID Project No. USAID RE III 388–0070. USAID: Dhaka, Bangladesh.

van der Hoek, EE and Comans, ENJ (1996) Modelling arsenic and selenium leaching from acidic fly ash by sorption on iron (hydr)oxide in the fly ash matrix. Env Sci Tech v.30, p.517–523.

Weigert, P., Muller, J., Klein, H., Zufelde, K.P. and Hillebrand, J. (1984) Arsen, Blei, Cadmium und Quecksilberin und auf Lebensmitteln. ZEBS Hefte, 1, Berlin.(In German).

Welch, A.H., Helsel, D.R., Focazio, M.J. and Watkins, S.A. (1999) Arsenic in ground water supplies of the United States. In: W.R. Chappell, C.O. Abernathy and R.L. Calderon (eds) Arsenic Exposure and Health Effects: proceedings of the Third International Conference on Arsenic Exposure and Health Effects, July 12–15, 1998, San Diego, California. p 9–17., Elsevier.

Welch, A.H., Lico, M.S. and Hughes, J.L. (1988) Arsenic in groundwater of the Western United States. Groundwater. v.26, p.334–347.

Yan Chu, Y (1994) In: J.O. Niragu, ed. Arsenic in the Environment, Part I: Cycling and Characterization. New York: John Wiley, p. 17–49.

Zheng,B., Yu,X., Zhand, J., and Zhou,D. (1996) Environmental geochemistry of coal and endemic arsenism in southwest Guizhou, P.R. China 30th International Geologic Congress. Abstracts v.3, p.410.

6

Geological Epidemiology: Coal Combustion in China

ROBERT B. FINKELMAN
U.S. Geological Survey, Reston, Virginia

HARVEY E. BELKIN
U.S. Geological Survey, Reston, Virginia

JOSE A. CENTENO
Armed Forces Institute of Pathology, Washington, D.C.

ZHENG BAOSHAN
Institute of Geochemistry, Guiyang, Guizhou Province, P.R. China

Introduction

In a recent report, the U.S. Environmental Protection Agency (EPA) has concluded that, with the possible exception of mercury, there is no compelling evidence that emissions from U.S. based coal-burning electric utility generators cause human health problems (EPA 1998). However, worldwide, the use of poor quality coal and/or the improper use of coal may cause or contribute to significant widespread human health problems. Health problems caused by impurities in coal, such as arsenic and fluorine, have been reported from the former Czechoslovakia (Bencko 1997) and from China. The World Bank (1992) estimates that between 400 and 700 million women and children are exposed to severe air pollution, generally from cooking fires. A substantial proportion of these people rely on coal for domestic cooking and heating and are thereby exposed to particulates, metal ions, gases (such as SO_x), and organic compounds causing potentially serious respiratory problems and toxic reactions.

Although addressing human health problems is the domain of biomedical and public health scientists, geoscientists have tools, skills, databases, and perspectives that may help the medical community address environmental health problems. Geoscientists are best equipped to characterize natural resources such as rocks, soils, and water. Various analytical tools used to characterize these natural materials have also been used effectively to characterize materials such as ambient dust and the products of coal combustion that cause or contribute to human health problems. In this chapter, we describe some of the geologic and geochemical tools being used to address arsenism and fluorosis caused by residential coal combustion in Guizhou Province, China.

Health Problems Caused by High Arsenic- and Fluorine-Bearing Coals

Wood had long been the primary energy source in southwest China, but by the early part of the twentieth century the forests were largely denuded and the residents were forced to use alternate sources of fuel. In southwest Guizhou Province, surface exposures of coal are plentiful and coal quickly became the primary fuel for domestic use. Unfortunately, some of these coals have undergone mineralization, causing their enrichment in potentially toxic trace elements such as arsenic, fluorine, mercury, antimony (Sb), and thallium (Tl).

Burning the mineralized coals in unvented stoves volatilizes toxic elements and exposes the local population to these emissions. The situation is exacerbated by the practice of drying crops directly over the coal fires. In the autumn, when it is commonly cool and damp in the higher elevations of Guizhou Province, the residents commonly dry their corn, chili peppers, and other foods directly over the burning coals.

Arsenic

Chronic arsenic poisoning, which affects at least 3000 people in Guizhou Province, has been described by Zheng et al. (1996). Those affected exhibit typical symptoms of arsenic poisoning including hyperpigmentation (flushed appearance, freckles), hyperkeratosis (scaly lesions on the skin, generally concentrated on the hands and feet), Bowen's disease (dark, horny, precancerous lesions of the skin: Fig. 6.1A), and squamous cell carcinoma.

Zheng et al. (1996) have shown that chili peppers dried over open coal-burning stoves may be a principal vector for the arsenic poisoning. Fresh chili peppers have less than one part-per-million (ppm) arsenic. In contrast, chili peppers dried over high-arsenic coal fires can have more than 500 ppm arsenic. Significant arsenic exposure may also come from other tainted foods, ingestion of dust (samples of kitchen dust contained as much as 3000 ppm arsenic), and from inhalation of indoor air polluted by arsenic derived from coal combustion. The arsenic content of drinking water samples was below the U.S. Environmental Protection Agency drinking water standard (1973) of 50 ppb and does not appear to be an important factor.

Recent research on the chemical and mineralogical characterization of arsenic-bearing coal samples from Guizhou Province (Belkin et al. 1997, 1998; Finkelman et al. 1999) has added to our knowledge of this problem. Instrumental neutron activation analyses of the coal indicate arsenic concentrations as high as 35,000 ppm! The extreme magnitude of this concentration can be seen by comparison with U.S. coals, where the mean arsenic concentration in nearly 10,000 U.S commercial coal samples is approximately 22 ppm, with a maximum value of about 2000 ppm (Bragg et al. 1997).

Belkin et al. (1997, 1998) examined polished blocks of the arsenic-rich Guizhou Province coal using a scanning electron microscope equipped with energy-dispersive X-ray analyzer (SEM–EDX), and an electron microprobe. This study also used X-ray diffraction analysis, optical microscopy, and other techniques to identify the arsenic minerals in the coal. A wide variety of As-bearing mineral phases was observed. Pyrite (FeS_2) is the most common sulfide. The range of As in pyrite determined by electron microprobe analyses is from the detection limit (~100 ppm) to about 4.5 weight percent. Arsenopyrite (FeAsS), a very rare mineral in coal in the U.S., is abundant in several of the Chinese coal samples. Other minor As-bearing minerals present include a third As-bearing sulfide, phosphates, sulfates, oxides, and silicates (clays).

The coal samples with arsenic concentrations in excess of 3 weight percent were mineralogically unusual. Although they contain small grains and framboids of pyrite, the concentration of arsenic in these forms of the iron sulfide phase is completely inadequate to account for the arsenic abundance on a

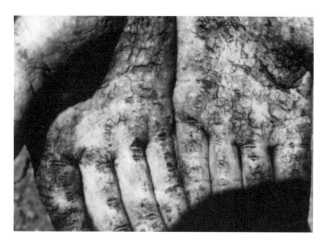

Figure 6.1: Extensive scaly lesions (hyperkeratosis) as evident on the hands of a resident of this region. Abundant cracks in the lesions offer access to pathogens as liver flukes, a major cause of death in this area.

bulk analysis of the coal. In SEM back-scattered electron images, a distinct banding characterized by different image brightness is observed (Figs. 6.2A, B). Semi-quantitative analysis by electron microprobe demonstrates that the bright bands contain ≈3 weight percent arsenic. No discrete As-bearing phase could be observed using electron microscopy at magnifications of 1 million times. Thus, finely-dispersed arsenopyrite, As-bearing pyrite, or any other As-mineral can be ruled out as the source of the arsenic. Using relatively new, sophisticated analytical procedures [X-ray absorption near-edge structure (XANES) and extended X-ray absorption fine structure (EXAFS)] (Huffman et al. 1994) it was determined that most of the arsenic in these samples is present as organically bound arsenate (AsO$_4^{3-}$).

The chemical and mineralogical characterization of the coals in Guizhou Province has had several benefits. Commonly coals containing high ash, high sulfur, and high contents of toxic elements such as arsenic are "cleaned" to reduce the concentrations of these undesirable components. The cleaning process physically separates the inorganic constituents from the organic constituents of the coal, generally resulting in a lower ash, lower sulfur, lower arsenic fuel. Because the arsenic in these special Guizhou Province coals is largely organically bound, cleaning would have produced a fuel even richer in arsenic. Information on the arsenic mineralogy may also help predict the behavior of arsenic during coal combustion. Preliminary characterization of residual ash in coal-burning stoves indicates high retention of arsenic. Mineralogical characterization in conjunction with combustion tests may determine if one or more of the arsenic-bearing phases is primarily responsible for adsorption of arsenic on the chili peppers.

Knowledge of the concentration of arsenic in these coals could be used to systematically map the distribution of this element and identify locations of high- and low-As coals. Understanding the relationship between the arsenic content of the coals and geologic processes could help to identify areas where similar situations may exist. Knowing where these high-As coals occur could result in measures to protect local populations from exposure to arsenic and the resultant health problems in Guizhou Province.

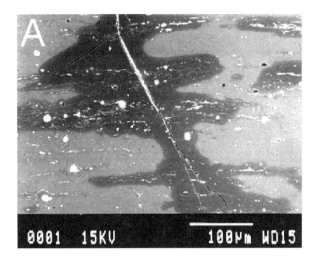

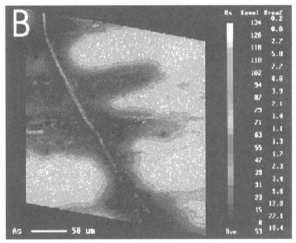

Figure 6.2:
A. SEM back-scattered electron image of polished block of arsenic-rich coal. Dark areas are coal, bright areas are mainly pyrite, milky area is coal containing organically bound arsenate. Fluids moving through the fracture in the coal appears to have removed arsenic from the organic matrix.
B. X-ray map depicting the distribution of arsenic in the coal. Light areas are high concentrations, and dark areas are low concentrations. Compare distribution of arsenic to the outline of the milky area in Figure 6.2A.

The tools and techniques employed to characterize the distribution of arsenic in these coals could be used to better understand the forms and behavior of other potentially toxic elements such as fluorine.

Fluorine

The health problems caused by fluorine volatilized during domestic coal use are far more extensive than those caused by arsenic. More than 10 million people in Guizhou Province and surrounding areas suffer from various forms of fluorosis (Zhang and Cao

DENTAL FLUOROSIS AND COAL–BEARING REGIONS OF CHINA

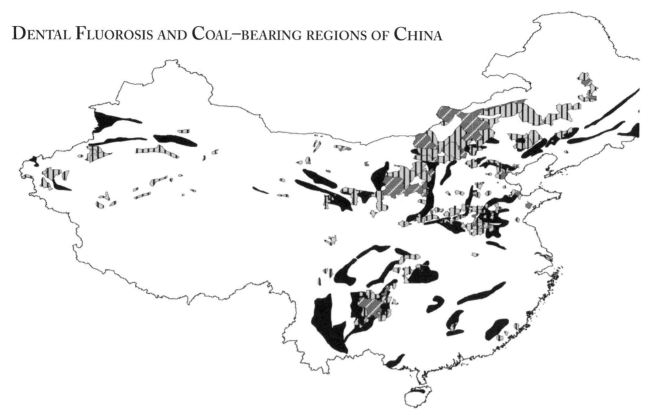

PREVALENCE RATES OF DENTAL FLUOROSIS

30–50% of Population
50–70% of Population
Coal-bearing Regions

Figure 6.3: Map of China produced by GIS techniques showing the distribution of coal-bearing regions. Superimposed on this are the areas of high incidence of dental fluorosis. The coincidence of dental fluorosis with Paleozoic coal-bearing rocks in the Guizhou Province and surrounding area is compatible with domestic coal combustion as the source of fluorine. The population burns coal briquettes composed of high F-bearing coal bound with high F-bearing clay as domestic fuel, compounding F exposure from high-F water in these areas. From the Open File Report #01-318 U.S. Geological Survey compiled by A.W. Karlsen, A.C. Schultz, P.D. Warwick, S.M. Podwysocki and V. S. Lovern, with permission.

1996, Zheng and Huang 1989). Fluorosis has also been reported in 13 other provinces, autonomous regions, and municipalities in China (Ando et al. 1998).

Typical symptoms of fluorosis include mottling of tooth enamel (dental fluorosis) and various forms of skeletal fluorosis including osteosclerosis, limited movement of the joints, and outward manifestations such as knock-knees, bow legs, and spinal curvature. Fluorosis combined with nutritional deficiencies in children can result in severe bone deformation (Fig. 6.4).

The etiology of fluorosis is similar to that of arsenism in that the disease is derived from foods dried

over coal-burning stoves. Ando et al. (1998) estimate that 97% of the fluoride exposure came from food consumption and 2% from direct inhalation.

Zheng and Huang (1989) have demonstrated that adsorption of fluorine by corn dried over unvented ovens burning high (>200 ppm) fluorine coal is the probable cause of the extensive dental and skeletal fluorosis in southwest China. The fluorine in the coal is primarily in the clay minerals (Belkin et al.1999). However, the fluorosis is compounded by the use of clay as a binder for making briquettes. The clay used is a high-fluorine (mean value of 903 ppm) residue formed by intense leaching of a limestone substrate. Chemical analysis can help to identify sources of low-fluorine coals and clays in the region so that low-fluorine briquettes can be produced.

Conclusion

Domestic coal use in developing countries can cause serious health problems because of the inclusion of potentially toxic elements in this fuel source. Geoscientists from other countries have the tools and the expertise to analyze, and forecast, the concentrations, distributions, and behavior of the harmful species. Collaboration between countries generating such immediate and useful information for a variety of circumstances would go a long way toward mitigating present and future health impacts on an unsuspecting public.

Figure 6.4: Bone and joint deformation due to nutritional deficiency combined with exposure to high levels of fluorine from residential coal combustion.

References

Ando, M., Tadano, M., Asanuma, S., Matsushima, S., Wanatabe, T., Kondo, T., Sakuai, S., Ji, R., Liang, C., and Cao, S., 1998, Health effects of indoor fluoride pollution from coal burning in China. Environmental Health Perspectives. v. 106, no. 5, p. 239–244.

Belkin, H. E., Zheng, B., Zhou, D., and Finkelman, R. B., 1997 Preliminary results on the geochemistry and mineralogy of arsenic in mineralized coals from endemic arsenosis areas in Guizhou Province, P.R. China. Fourteenth Annual International Pittsburgh Coal Conference CD–ROM.

Belkin, H.E., Warwick, P., Zheng, B., Zhou, D., and Finkelman, R. B., 1998, High arsenic coals related to sedimentary rock–hosted gold deposition in southwestern Guizhou Province, People's Republic of China. Fifteenth Annual International Pittsburgh Coal Conference CD–ROM.

Belkin, H. E., Finkelman, R. B., and Zheng, B.S., 1999, Geochemistry of fluoride–rich coal related to endemic fluorosis in Guizhou Province, China. Pan–Asia Pacific Conference on Fluoride and Arsenic Research. Abstract 45, p. 47.

Bencko, V., 1997, Health aspects of burning coal with a high arsenic content: the Central Slovakia experience, in Arsenic: Exposure and Health Effects, chapter 8, C.O. Abernathy and R. L. Calderon, eds., Chapman and Hall, New York, p. 84–92.

Bragg, L. J., Oman, J. K., Tewalt, S. J., Oman, C. L., Rega, N. H., Washington, P. M., and Finkelman, R. B., 1997, U.S. Geological Survey Coal Quality (COALQUAL) Database: Version 2.0. U.S. Geological Survey Open–File Report 97–134 (CD–ROM).

Finkelman, R. B., Belkin, H. E., and Zheng, B., 1999, Health impacts of domestic coal use in China. Proceedings National Academy of Science, U.S.. Vol. 96, p.3427–3431.

Huffman, G. P., Huggins, F. E., Shah, N., and Zhao, J., 1994, Speciation of arsenic and chromium in coal and combustion ash by XAFS spectroscopy. Fuel Processing Technology, v. 39, no. 1–3, p. 47– 62.

USEPA (U.S. Environmental Protection Agency), 1973, Water Quality Criteria 1972, EPA R3 73033. Government Printing Office, Washington, D.C.

USEPA (U.S. Environmental Protection Agency), 1998, Mercury study report to Congress: White paper. (http:www.epa.gov/oar/merwhite.html) February 24, 1998.

World Bank, 1992, The World Bank Development Report 1992: Development and the Environment, Washington, D.C., p.53.

Zhang, Y. and Cao, S. R., 1996, Coal burning induced endemic fluorosis in China. Fluoride, v. 29, no. 4, p. 207–211.

Zheng, B. and Huang, R., 1989, Human fluorosis and environmental geochemistry in southwest China. Developments in Geoscience, Contributions to 28th International Geologic Congress, 1989, Washington, D.C. Science Press, Beijing, China. p. 171–176.

Zheng, B., Yu, X., Zhand, J., and Zhou, D., 1996, Environmental geochemistry of coal and endemic arsenism in southwest Guizhou, P.R. China. 30th International Geologic Congress Abstracts, Vol. 3, p. 410.

7

Mitigation of Endemic Arsenocosis with Selenium: An Example from China

WANG WUYI, YANG LINSHENG, HOU SHAOFAN, AND TAN JIAN'AN
Institute of Geographical Sciences and Natural Resources Research
Chinese Academy of Sciences, Beijing, P.R. China

Introduction

Endemic arsenocosis (chronic arsenic poisoning) in China comes from two sources of arsenic (As). One source is drinking water, with As concentrations 2–40 times that of the state standard of 0.05 mg/l As. The second is smoke pollution from combustion of coal with high concentrations of As; this can be inhaled or ingested from smoke-contaminated food. Over 2,000,000 people live in areas of high geological As concentrations (Cao 1996), and more than 17,000 arsenocosis patients in 21 counties of five provinces or Autonomous Regions (Fig. 7.1) have been identified.

Long-term exposure to As in air, diet, or drinking water can result in permanent and severe damage to health, including lesions of the skin, mucous membranes of the digestive, respiratory, circulatory, and nervous systems, and rhagades (skin cleft on palm and feet). Elevated As intake is also associated with skin, liver, and lung cancers (Centeno 2000, Liang 1999, Wang Lianfang 54–61 1997).

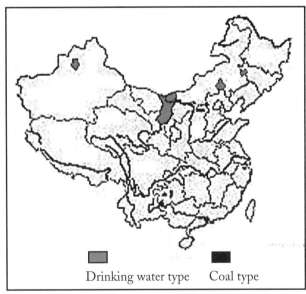

Figure 7.1: Geographic distribution of two sources of As causing endemic arsenocosis in China.

At present, there are few studies of efficient measurement of treatment of endemic arsenocosis patients. Our study demonstrates that treatment of these patients

51

with dietary selenium (Se) can cause both excretion (elimination) of As accumulated in the human body and remediation of some health damages. We report the results of this experiment.

Methods

Data were collected on 3 test groups of people: 186 patients, from BaYinMaoDao Farm in Inner Mongolia suffering from endemic arsenocosis, were divided into a treatment group (100 patients) and a control group (86 patients). A third group, consisting of 70 families, received no treatment but drank ambient well water, >0.10 mg/l As. All participants had been exposed to high-As drinking water (>0.10 mg/l) since 1983. Throughout the experiment, water containing 0.05 mg/l As was supplied for both treatment and control groups. Of the 186 patients, 100 were treated with Se-enriched yeast* tablets, containing 100 μg Se/tablet. The treatment lasted 14 months. Treated patients received 100–200 μg Se/day.

All patients were examined for clinical criteria of arsenocosis: characteristic pigmentation, depigmentation, hyperkeratosis, rhagades (skin cleft), and incidence of secondary symptoms of headaches, dizziness, thoracalgia (chest pain), numbness of hands or feet, convulsions, or lumbago. Tests on liver function, liver ultrasonotomography, electrocardiography, and electron microscope observation of erythrocytes were performed. Doctors followed the Standard of Chinese Endemic Arsenocosis Clinical Diagnosis Guidelines.

Hair, urine, and blood samples were collected before the experiment and at the end of the 3rd, 6th, 9th, and 14th month. Assay results from the initial and final samples are presented.

Human hair (255 samples) was collected; about 5 g of new growth hair was cut from patients in both the treatment and control groups and the third, ambient well water, group over the 14-month period. Stainless steel scissors were used. Before analysis, the hair samples were dipped in neutral detergent, washed with running water, distilled water, and ion–free water in turn, then oven-dried for 4 hours at 60° C, and cut into 0.5 cm segments for digestion. Urine samples were collected at the same time. Samples of drinking well water for the 70 families in the high-As region were collected; 100 ml water were directly taken from each well, stored in acid-washed plastic bottles and kept refrigerated (4°C). All the samples were analyzed within 1 week.

All water, hair, and urine samples were analyzed by hydride generation coupled with ICP–AES. Samples (0.3 g hair; 5 ml water; 5 ml urine) were digested with concentrated nitric acid (3 ml) and perchloride acid (1 ml) by electrothermal heating until the perchloride acid was almost driven off. After the samples were cooled, 2 ml hydrochloric acid were added and made to standard volume of 8 ml. Samples were analyzed by hydride generation inductively coupled plasma atomic emission spectrometry

Table 7.1: Endemic arsenocosis incidence and As in drinking water			
Water As μg/ml	*Population*	*Patients identified*	*%*
<0.05	182	0	0
0.05~0.10	340	33	9.71
>0.10	1480	278	18.78
Total	2002	311	15.53

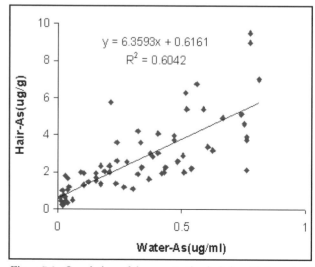

Figure 7.2: Correlations of As concentration in hair with As content in drinking water (r = 0.777, n = 70, p < 0.05)

* Manufactured by the Beijing Tiancifu Medicine Company

(ICP2070, Baird Co.). ICP2070 has a very sensitive DL (detection limit) at 0.8–1.6 mg/ml and RSD (relative standard deviation) of 1.62–2.71%. Instruments may have slightly different DL due to changes in experimental conditions. For quality control in chemical analysis, two standard reference samples (Chinese hair, GBW0901, As content: 0.59±007 µg/g, Chinese Standard Sample Study Center, Chinese Academy of Measurement Sciences) were randomly analyzed with each batch of test hair samples (Wang Lizhen).

Results and Discussion

1. Incidence of arsenocosis increased with As concentration of drinking water (Table 7.1), suggesting endemic arsenocosis can result from exposure to As in drinking water.

2. Hair samples from families drinking ambient well water shows that hair-As content of adults in arsenocosis-affected regions has a significant positive correlation with As content in drinking water ($r = 0.777, n = 70, p < 0.05$) (Fig. 7.2). Smith (Smith 1064) studied 1250 hair-As samples in England. He determined that if As in 80% of hair samples was less than 1.0 µg/g, the mean was 0.81 µg/g and the median was 0.51 µg/g. He concluded that if hair-As is lower than 2.0 µg/g, arsenocosis has not occurred; if hair-As is 2.0–3.0 µg/g, the people should be evaluated; if hair–As is higher than 3.0 µg/g it is abnormally high and an indication of possible arsenocosis. Our work shows that hair-As is a good indicator of As levels in both the human body and the environment.

3. A preliminary clinical examination was carried out according to the Criteria for Clinical Manifestations for Arsenocosis (Table 7.2) (Wang Lianfang 184–191, 1997).

(Note: Symptoms in nerve-blood vessels include three syndromes: neurasthenia, peripheral neuritis, and Raynaud's disease. These medical terms indicate abnormalities in the nervous system. Keratosis is a general term indicating keratinization and abnormal squamous cell development.)

Results indicate that after 14 months, 75.0% of Se-treated patients recovered clinically to some extent and 55.0% had decreased secondary symptoms, but only about 25.6% and 21.4%, respectively, of the control group (Table 7.3). In contrast, the unvaried cases and deterioration for the control group are higher than Se-treated group.

4. Before visible changes on the skin appear, some biophysical–biochemical or/and physiological–patho-

Table 7.2: Chinese clinical criteria for arsenocosis

	Grade I	Grade II	Grade III
Keratosis	Mild hyper-pigmentation	Papular eruptions	Lesions on palms metatarses, backs of hands and feet
Pigmentation	Grayish or black, or brown spots	Brown–gray with increased brown spots	Grayish black or dense brown spots
Depigmentation	Scattered	More spots	Densely clustered spots
Nerve-blood vessel symptoms	1 item	2 items	3 items

Table 7.3: Comparison between Se-therapy group and control group by the clinical exam for arsenocosis

	Recovery Case %		Unvaried Case %		Deterioration Case %	
Se-therapy group						
Clinical exam	75	75.0	21	21.0	1	1.0
Symptom	55	55.0	45	45.0	0	0
Control group						
Clinical exam	22	25.58	52	60.47	12	13.96
Symptom	21	21.41	51	59.30	14	16.24

Recovery = symptoms and signs disappear or are alleviated.
Unvaried = no significant change for symptoms and signs, or the change <one grade.
Deterioration = worsening more than one grade or results in Bowen's disease (7,4).
Clinical exam, see Table 7.2.
Symptom = secondary symptoms, e.g., headache.

logical changes can be detected such as hepatomegaly (liver swelling); disorder of heart and other disorders that can be detected by a check of liver function; liver ultrasonotomography (ultrasonic equipment for medical diagnosis); electrocardiography; and electron microscope observation of erythrocytes. Patients' erythrocytes change morphologically; target cells, spur cells, echinocytes, and spherocytes can be seen in very high ratio. With supplemental Se, these abnormalities can be remedied to normal. In the Se-therapy group, liver function, liver supersonic tomography, electrocardiography, and electron microscope observation of erythrocytes reversed significantly compared to the control (Table 7.4). This means that not only can Se protect erythrocytes and remediate them from lesions, but also that this information provides an important contribution toward understanding the mechanism of As-induced lesions.

5. Data from Table 7.5 indicate that both the Se-therapy group and the control group have a significant decrease of As concentration in blood, urine and hair, but As concentrations of Se-therapy group decrease much more than that of control group.

Selenium is a nutritionally required element. It has been known as an antagonist of arsenic toxicity for many years (Levander 1966). Recent studies showed that selenium could form a compound with glutathione

and arsenic in rabbit bile after injecting intravenously with sodium selenite followed immediately with intravenous sodium arsenite (Aposhian 1999).

Cessation of chronic As exposure can reduce the As concentration in urine, but longer periods are required to restore to normal background urinary excretion (Buchet al. 1999). In our study, urinary As concentration in the control group decreased after exposure to elevated As in drinking water ceased.

Cessation of elevated As exposure can also reduce As in hair. But As concentration in hair persists even after use of the polluted well water ceased for 2–4 months (Maki-Paakkanen 1998), and even after 2.5 years (Li Yong 1998). However, the differences in our study of the decreasing concentration and decreasing rate between the Se-therapy group and control group show that Se has an efficient effect on the binding of As in hair.

Many areas of the world where chronic arsenic exposure occurs are low in selenium (Aposhian 1998), and the results in this study show that selenium supplementation can effectively decrease arsenic concentrations in hair, urine, and blood. This study is among the first indicating the mitigation of arsenocosis by dietary selenium supplementation. Tests should be carefully carried out on the basis of evaluation of selenium intakes or nutritional status of residents in other areas.

Table 7.4: Comparison of physical and chemical test on Se group and control

	Liver Function	Liver UT (a)	EMOE(b)	Electro-cardiography
Se-therapy Group	80.00%	60.0%	72.22%	84.78%
Control group	46.15%	30.7%	0	44.83%

(a) Liver ultrasonotomorgraphy: ultrasonic equipment for medical diagnosis on liver.
(b) Electron microscope observation of erythrocyte.

Table 7.5: As in blood, urine and hair before and after treatment with Se

		Before treatment	After treatment
Se-therapy group	Blood As µg/ml	0.070±0.099 (98) (a)	0.016±0.006(88) (b)
	Urine As µg/ml	0.255±0.306 (99) (b)	0.030±0.030 (85)
	Hair As µg/g	2.970±1.627 (99)	0.798±0.603 (88) (b)

(a) Data in bracket is sample number.
(b) P <0.05 between Se-therapy group and control

Acknowledgement

We wish to express our thanks to the Chinese Academy of Sciences supporting this study of projects KZ–951–B1–204 and CXIOG–A00–01.

References

Aposhian H.V., Zakharyan R.A., Wildfang E.K., Healy S.M., Gailer J., Radabaugh T.R., Bogdan G.M., Powell L.A., Aposhian M.M., 1999, in Arsenic Exposure and Health Effects, Proceedings of the Third International Conference on Arsenic Exposure and Health Effects, W.R. Chappell, C.O. Abernathy and R.L. Calderon, eds, July 12–15, 1998, San Diego, California. Elsevier Science, Amsterdam, the Netherlands, p.289–297.

Aposhian H.V., et al. (more than 16 co–authors), 1999, DMPS–arsenic challenge test II.modulation of arsenic species, including monomethylarsonous acid (MMAIII), excreted in human urine. Toxicology and Applied Pharmacology, v. 165, p.74–83.

Buchet J.P., Hoet P., Haufroid V., and Lison D., 1999, Consistency of biomarkers of exposure to inorganic arsenic: Review of recent data, in Arsenic Exposure and Health Effects, Proceedings of the Third International Conference on Arsenic Exposure and Health Effects, W.R. Chappell, C.O. Abernathy and R.L. Calderon, eds, July 12–15, 1998, San Diego, California. Elsevier Science, Amsterdam, the Netherlands, p.31–40.

Cao Shouren, 1996, Status of investigation and study on inorganic arsenic pollution in China. Chinese Journal of Endemiology, v. 5 (supplement), p. 1–4. (in Chinese with English abstract).

Centeno, J.A., et al., 2000, Arsenic–induced lesions: Syllabus, Armed Forces Institute of Pathology and American Registry of Pathology, 46p.

Levander O. A., and Baumann C. A., 1966, Selenium metabolism. VI. Effects of arsenic on the excretion of selenium in the bile. Toxicology and Applied Pharmacology, v. 9, p.106–115.

Li Yong, Gao Biyu, Wang Guoquan, 1998, Clinical tracing investigation of 518 endemic patients. Journal of Environment and Health, v. 8, no. 4, p. 152–155 (Chinese with English Abstract).

Liang Xiufen, Dai Qin et al., 1999, A study on the relationship of high–arsenic in drinking water to lung cancer, Chinese Journal of Prevention and Control of Chronic Non–communicable Diseases, v. 7, no. 2, p. 1–4 (in Chinese with English abstract).

Maki–Paakkanen J., Kurttio P., Paldy A., Pekkanen J., 1998, Association between the clastogenic effect in peripheral lymphocytes and human exposure to arsenic through drinking water. Environmental and Molecular Mutagenesis, v. 32, p. 301–313.

Smith H., 1964, Interpretation of the arsenic content of human hair. Forensic Sci. Soc., v. 4, no. 4, n. 192–199.

Wang Lianfang, 1997, Endemic arsenism and black foot disease, Xinjiang Science and Public Health Press, p.54–61, p.184–191 (in Chinese).

Wang Lizhen, Hou Shaofan, Yang Lingheng, 1999, determination of arsenic in hair, blood and urine by ICP–AES equipment with hydride generator. Chinese Journal of Spectroscopy Laboratory, v. 16, no. 4, p. 385–387 (in Chinese with English abstract).

8

Biogeochemical Cycling of Iodine and Selenium and Potential Geomedical Relevance

EILIV STEINNES
Department of Chemistry
Norwegian University of Science and Technology, Trondheim, Norway

Introduction

An increasing number of the 92 naturally occurring elements on Earth are now known to be essential to humans and other vertebrate species. In addition to the ten main constituents (H, N, O, Na, Mg, P, S, Cl, K, Ca), twelve elements present in trace quantities are generally accepted to have necessary functions in the human body. These essential trace elements are Mn, Cr, Fe, Co, Cu, Zn, Se, Mo, F, bromine (Br), Si, and I. Most of these elements are present in human serum at levels orders of magnitude lower than their mean concentrations in the Earth's crust (e.g., Mn and Cr are less than 10^{-6} of average rock composition), except for I and Se, which occur in similar concentrations in human serum and in rocks. This indicates that the pathways of these two elements to humans are basically different from those of other essential trace elements. There is now substantial evidence to suggest that the marine environment is an important source of I and Se to humans and other terrestrial species through biogeochemical cycling involving atmospheric transport.

The Ocean as a Source of Elements to Humans and Livestock

The dissolved matter in ocean water is enriched relative to the earth's crust in a few elements (e.g., Na, Mg, S, Cl, Br) but depleted in most others. Some elements, such as I and Se, are strongly enriched in marine organisms relative to their concentrations in sea water. Fish and other marine food may constitute an important source of these elements to humans. It has become increasingly evident, however, that atmospheric transport of substances from marine to terrestrial systems may constitute an alternative pathway of some essential elements to humans and livestock.

Elements may be released from the ocean surface as:

- Seasalt aerosols (Na^+, Mg^{2+}, Cl^-, SO_4^{2-}, etc.)
- Biogenic gases ($(CH_3)_2S$, CH_3I, etc.)

In addition, some elements may be enriched in surface-active material present on the surface of the ocean (Dean 1963) either from atmospheric deposition of particulate material or from enrichment in marine biota from which the material is derived. When bubbles burst on the ocean surface, this micro-layer may be enriched on salt particles formed and thus be preferentially released to the atmosphere.

The importance of the marine environment as a source of cation supply to coastal areas in Norway was convincingly demonstrated in a study of forest soils by Låg (1968). Later work in Norway on natural surface soils (Njåstad et al.1994) and lake waters (Allen and Steinnes 1986) has confirmed that atmospheric cycling processes play an important role in supplying elements such as Na and Mg to the terrestrial surface in coastal regions. Similarly, the concentrations of halogens (Cl, Br, I) in natural surface soils were shown to be several times higher near the coast of mid-Norway than in areas east of the Caledonian Mountain Range where the marine influence is small (Låg and Steinnes 1976). A similar geographical trend was shown for Se, which is present in the ocean at a level of only 0.1 μg L^{-1} (Låg and Steinnes 1974, 1978), and it was hypothesized that there might be a marine source for the excess Se in the coastal soils.

In the following, the present knowledge on the biogeochemical cycling of I and Se in relation to the marine environment is briefly reviewed, and some possible geomedical consequences are indicated.

Iodine

It has been obvious for a long time that iodine must be released from the ocean in another form than seasalt aerosols. Whereas the Cl/I ratio in ocean water is about 3×10^5, it is generally 100–1000 times lower in precipitation samples collected in marine air (Seto and Duce 1972) and another factor of 10 lower in natural surface soils (Låg and Steinnes 1976). Different hypotheses for the nature of the separation process causing iodine enrichment in the marine atmosphere relative to the sea water have been presented. Miyake and Tsunogai (1963) proposed a photochemical oxidation of iodide to elemental iodine in the surface layer of the sea. On the other hand, an observation that the iodine in the atmosphere is to a great extent present in organic form (H.I. Svensson and E. Eriksson, cited in Bolin 1959) led Dean (1963) to suggest that iodine may be enriched in the organic surface microlayer and correspondingly enriched on the salt particles formed when bubbles break through the surface. More recent research has indicated that the main source of iodine in the marine atmosphere may be CH_3I produced by biological activity in the surface water (Whitehead 1984).

Selenium

As far as is known, the first indication that the ocean might be a source of selenium to the terrestrial environment came from an investigation by Låg and Steinnes (1974 1978). In a study on the geographical distribution of trace elements in the humus layer of forest soils, they found that the Se concentration near the coast of mid-Norway was as much as 5 times higher (1.0 versus 0.2 ppm) than in areas of eastern Norway shielded from oceanic influence by high mountains. Låg and Steinnes (1974) put forward the hypothesis that the excess Se observed in the coastal surface soil may have been supplied by precipitation, with the marine environment as a source.

At first sight, this hypothesis might seem rather unlikely since the Se concentration in ocean water is very low, about 0.1 μg L^{-1} (Cutter and Bruland 1984). Further studies in Norway on natural soils (Steinnes 1991), agricultural soils (Wu and Låg 1988), and ombrogenous peat (Steinnes 1997) and in Sweden on forest soils (Johnsson 1989) verified and reinforced the original findings by Låg and Steinnes (1974). Studies of atmospheric deposition of Se based on analysis of terrestrial mosses (e.g., Steinnes et al. 1992) confirmed a strong coast-inland gradient, as illustrated in Figure 8.1. Investigations of Se in peat cores from ombrotrophic bogs showed that the same trend was evident before the Industrial Revolution (Steinnes 1997). Although the selenium in the sea water could hardly be a source of excess Se in coastal areas as such, the possibility of a biologically driven cycling of Se, in a similar way as indicated above for I, could not be excluded. It was already known that Se is enriched in marine food chains (Lunde 1968).

Thus, there is strong evidence from research carried out in Norway that the ocean is a significant source of Se to coastal terrestrial areas, presumably through biogeochemical cycling processes. More recent evidence from marine studies appears to confirm this hypothesis:

- Cutter and Bruland (1984) determined the concentration level of Se in the ocean to be of the order of 0.1 μg^{-1}, and that organic selenide made up around 80 % of total dissolved Se in ocean surface waters.

- Mosher et al. (1987) observed anomalous enrichment of Se in marine aerosols, and found that the Se concentration was related to primary productivity in the sea.

• Cooke and Bruland (1987) studied the speciation of dissolved Se in surface water, and observed the formation of volatile organoselenium compounds, mainly dimethyl selenide, $(CH_3)_2Se$. On that basis they hypothesized that outgassing of dimethyl selenide may be an important removal mechanism for dissolved Se from aquatic systems.

• Amouroux and Donard (1997) observed emission of Se to the atmosphere via biomethylation processes in the Gironde estuary, France.

Altogether, these observations appear to confirm that biological processes in the surface ocean water provide a dominant source of atmospheric Se supply to coastal regions.

Still, one important question remains to be answered in order to close the natural biogeochemical selenium cycle: What are the chemical transformations of Se when it leaves the ocean until it is supplied to the terrestrial surface and eventually is bound in the surface soil? This knowledge is also relevant with respect to another important issue: to what extent does Se derived from the marine environment enter terrestrial food chains? More research needs to be done to answer these questions.

Potential Geomedical Relevance

An extensive discussion of the geomedical importance of the above findings is beyond the scope of this chapter. From the literature, however, there are numerous examples that deficiency problems related to I and Ss are evident in many populations around the world (Låg 1990). These problems are more prominent in certain geographical areas than in others, and often related to predominant consumption of locally derived food.

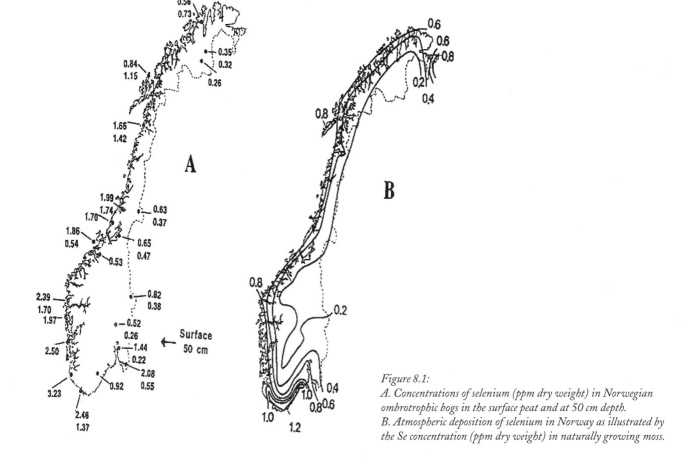

Figure 8.1:

A. Concentrations of selenium (ppm dry weight) in Norwegian ombrotrophic bogs in the surface peat and at 50 cm depth.
B. Atmospheric deposition of selenium in Norway as illustrated by the Se concentration (ppm dry weight) in naturally growing moss.

For example, some populations may live in territories where the bedrock contains too little quantities of I and Se. The soil in these areas may not provide sufficient amounts of these elements to prevent deficiency problems in the people, and in their domestic animals, if they are dependent on locally grown plants. Problems related to I and Se deficiency may therefore be more widespread, especially in developing countries, than has been so far anticipated.

Conclusions

Iodine and selenium are released from the ocean surface through volatile organic substances resulting from biogenic processes in the water. These elements are also enriched in surface soils, peats, and lake waters in coastal areas of Norway relative to corresponding areas farther inland. On the basis of these observations and evidence from international marine research, it is argued that biogeochemical cycling of I and Se from the ocean may be a significant source of these elements to the coastal terrestrial environment in general. Since these biologically essential elements often occur in very low concentrations in crustal rocks, the atmospheric supply of I and Se could be a significant geomedical factor in many areas of the world, and deficiency problems would most likely occur in regions far from the ocean.

References

Allen, R.O. and Steinnes, E., 1986. A contribution to the geochemistry of lakes in Norway. NGU Bulletin, v.409, p.35–48.

Amouroux, D. and Donard, O.F.X., 1997. Evasion of selenium to the atmosphere via biomethylation processes in the Gironde estuary, France. Marine Chem. v.58, p.173–188.

Bolin, B., 1959. Note on the exchange of iodine between the atmosphere, land and sea. Int. J. Air Poll. v.2, p.127–131.

Cooke, T.D. and Bruland, K.W., 1987. Aquatic chemistry of selenium: Evidence of biomethylation. Environ. Sci. Technol. v.21, p.1214–1219.

Cutter, G.A. and Bruland, K.W., 1984. The marine biogeochemistry of selenium: A re-evaluation. Limnol. Oceanogr. v.29, p.1179–1192.

Dean, G.A., 1963. The iodine content of some New Zealand drinking waters with a note on the contribution from sea spray to the iodine in rain. N. Z. J. Sci. v.6, p.208–214.

Johnsson, L., 1989. Se levels in the mor layer of Swedish forest soils. Swed. J. Agric. Res. v.19, p.21–28.

Låg, J., 1968. Relationships between the chemical composition of the precipitation and the content of exchangeable ions in the humus layer of natural soils. Acta Agric. Scand. v.18, p.148–152.

Låg, J., ed., 1990. Geomedicine. CRC Press, Boca Raton. 278 pp.

Låg, J. and Steinnes, E., 1974. Soil selenium in relation to precipitation. Ambio v.3, p.237–238.

Låg, J. and Steinnes, E., 1976. Regional distribution of halogens in Norwegian forest soils. Geoderma v.16, p.317–325.

Låg, J. and Steinnes, E., 1978. Regional distribution of selenium and arsenic in humus layers of Norwegian forest soils. Geoderma v.20, p.3–14.

Lunde, G., 1968. Activation analysis of trace elements in fishmeal. J. Sci. Fd. Agric. v.19, p.432–434.

Miyake, Y. and Tsunogai, S., 1963. Evaporation of iodine from the ocean. J. Geophys. Res. v.68, p.3989–3993.

Mosher, B.W., Duce, R.A., Prospero, J.M. and Savoie, D.L., 1987. Atmospheric selenium: Geographical distribution and ocean to atmosphere flux in the Pacific. J. Geophys. Res. v.92, p.13277–13287.

Njåstad, O., Steinnes, E., Bølviken, B. and Ødegård, M., 1994. National survey of element composition in natural soil. Results for samples collected in 1977 and 1985 obtained by ICP emission spectrometry. Report NGU 94.027, Geological Survey of Norway, Trondheim, 113pp (In Norwegian).

Seto, F.Y.B. and Duce, R.A., 1972. A laboratory study of iodine enrichment on atmospheric sea-salt particles produced by bubbles. J. Geophys. Res. v.77, p.5339–5349.

Steinnes, E., 1991. Influence of atmospheric deposition on the supply and mobility of selenium and cadmium in the natural environment. In Låg, J., Ed., Human and animal health in relation to circulation processes of selenium and cadmium, pp. 137–152.

Steinnes, E., 1997. Trace element profiles in ombrogenous peat cores from Norway: evidence of long range atmospheric transport. Water, Air, Soil Pollut. v.100, p.405–413.

Steinnes, E., Rambæk, J.P. and Hanssen, J.E., 1992. Large scale multi-element survey of atmospheric deposition using naturally growing moss as biomonitor. Chemosphere v.25, p.735–752.

Whitehead, D.C., 1984. The distribution and transformation of iodine in the environment. Environ. Internat. v.10, p.321–339.

Wu, X. and Låg, J., 1988. Selenium in Norwegian farmland soils. Acta Agric. Scand. v.38, p.271–276.

9

Environmental Iodine in Iodine Deficiency Disorders, with a Sri Lankan Example

FIONA M. FORDYCE
British Geological Survey, West Mains Road, Edinburgh, United Kingdom

CHRIS C. JOHNSON
British Geological Survey, Keyworth, Nottingham, United Kingdom

CHANDRA B. DISSANAYAKE AND UDAYA R. B. NAVARATNE
Dept. of Geology, University of Peradenyia, Kandy, Sri Lanka

Iodine Deficiency Disorders

Iodine is an essential element for human and other animal health and forms an important constituent of the thyroid hormones thyroxine (T4, also known as tetraiodothyronine) and tri-iodothyronine (T3). These hormones play a fundamental biological role in controlling growth and development (Hetzel and Maberly 1986).

If the amount of utilizable iodine reaching the thyroid gland is inadequate, or if thyroid function is impaired, hormone production can be reduced, resulting in a group of conditions collectively referred to as Iodine Deficiency Disorders (IDD) (Fernando et al. 1987, Hetzel 1989). The World Health Organization (WHO 1993) estimate that in excess of one billion people worldwide are at risk from IDD, the most common manifestation of which is goiter (Fig. 9.1). Iodine deficiency is the world's most common cause of preventable mental retardation and brain damage, and has a significant negative impact on the social and economic development of communities.

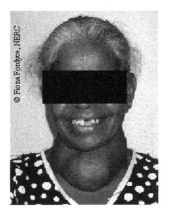

Figure 9.1: Sri Lankan woman suffering from the iodine deficiency disorder goiter.

Although it is likely that IDD are multifactorial diseases involving other trace element deficiencies and goitrogens (goiter-promoting substances) in foodstuffs, a lack of adequate dietary iodine remains a major concern (Stewart and Pharaoh 1996). The link between environmental iodine and IDD has been known for the last 80 years. During this time, the medical community has become well organized when tackling the problem, exemplified by the work of the International Council for the Control of IDD (ICCIDD) (<http://www.tulane.edu/~icec/icciddhome.htm>), which provides an excellent dissemination point for discussion and information. Remediation strategies often focus on enhancing dietary intakes of iodine via the introduction of iodinated salt and iodinated oil programs (Stanbury and Hetzel 1980). However, these methods are not always successful and other strategies, including environmental interventions, require development (DeLong et al. 1997). In contrast to the wealth of information about the symptoms, assessment, and treatment of IDD, there is very little on the primary cause,

61

a lack of readily available iodine in the environment and diet.

Environmental Iodine

Our knowledge of environmental iodine geochemistry is limited, mainly because the analytical methods for assessment are not routine and iodine is not an element that has been systematically determined in geochemical surveys. However, in the past two decades, improved analytical methodologies and an interest in iodine from different perspectives have added much to our knowledge. Exploration geochemists have used iodine as a pathfinder element to locate deeply buried mineralization, increasing an understanding of iodine movement in rocks and soils (Fuge et al. 1986). Environmental scientists have demonstrated the importance of atmospheric cycling of this element from the oceans (e.g., Alicke et al. 1999). More recently, research into the behavior of iodine in the environment has been connected to the nuclear industry and the threat posed by radionuclides of iodine. In the aftermath of nuclear accidents, ^{131}I readily finds its way through the food chain to humans where it is preferentially concentrated in the thyroid and may lead to thyroid cancer (Tuttle and Becker 2000). Research in this field has led to a much better understanding of the migration of iodine in the environment. In particular, soil fixation and volatilization to the atmosphere from the soil-plant interface are both far more significant in the geochemical cycle of iodine than previously recognized (Muramatsu et al. 1995, Schmidtz and Aumann 1995).

Despite these recent advances, there are myths surrounding environmental iodine that are perpetuated in the literature. Glaciated soils are often quoted as low in iodine, although there is no real evidence to support this, and communities in remote highland regions are commonly cited as being most at risk from IDD (WHO 1993). Whether this is due to the remoteness or to environmental factors has not been established (Stewart and Pharaoh 1996).

While it is a well-known fact that the oceans constitute major environmental sinks for iodine, which is volatilized from sea water and deposited on land during precipitation, investigators often assume an inverse linear relationship between iodine concentrations and distance from the sea (**Steinnes**). However, there is

growing evidence to suggest that the mechanisms of iodine volatilization and transport are complex (Fuge 1996).

Iodine and Goiter in Sri Lanka

A case in point is the island of Sri Lanka. Located in the Indian Ocean, no part of the island is more than 110 km from the sea, and yet endemic goiter [estimated to affect 10 million people (Fernando 1987)] has been recorded for the past 50 years. Goiter prevalence closely follows the climato-topographic regions of Sri Lanka and is a greater problem in the Wet Zone in the center and southwest of the island than in the Dry Zone to the north (Fig. 9.2). Interestingly, the southwest coastal region has some of the highest prevalence of goiter (Fordyce et al. 1998). Previous investigators had suggested that iodine was washed out of the soil by high rainfall in the Wet Zone, hence the high goiter prevalence in this region (Mahadeva and Shanmuganathan 1967). Although iodinated salt programs have been introduced in Sri Lanka, these have been only partially successful due to poor uptake (Dr. A. B. C. Amarasinghe, personal communication).

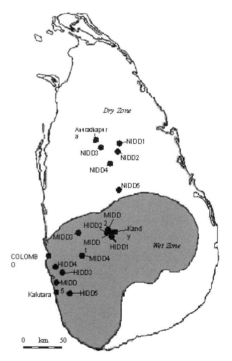

Figure 9.2: Sketch map showing the location of the 15 study villages; the Wet Zone/ Dry Zone demarcation used in the present study is based on the 2000-mm isohyet. The three groups of goiter prevalence villages are as follows: NIDD = No/low < 10% goiter prevalence; MIDD = Moderate 10–25% goiter prevalence and HIDD = High > 25% goiter prevalence.

In a project to investigate the selenium and iodine status of the environment and population of Sri Lanka, the present investigators examined the relationships between soil geochemistry and rice, the staple food crop of Sri Lanka. Soil and rice samples (*n* = 75) were collected from 15 villages, 5 in each of three goiter prevalence areas: low (NIDD) < 10%; moderate (MIDD) 10–25%; and high (HIDD) > 25% (Fig. 9.2). Total iodine concentrations were determined in soils by an automated colorimetric method (Fuge et al. 1978) and in 15 composite rice samples (1 composite per village) by epithermal Neutron Activation Analysis at the Environmental Analysis Section, Imperial College Centre for Environmental Technology, Silwood Park, U.K..

Results demonstrated that, contrary to popular belief, the concentrations of soil total iodine in the Wet (HIDD and MIDD) and Dry Zones (NIDD) of Sri Lanka are similar (Fig. 9.3) and are no lower than in soils from other parts of the world where goiter is not prevalent (Fordyce et al. 1998). However, further investigations into the soil geochemistry revealed that soils in the HIDD and MIDD goiter villages had higher organic matter, gibbsite, and goethite contents and lower pH than soils in the nongoiter (NIDD) villages. Thus, iodine in the Wet Zone (MIDD and HIDD) is adsorbed onto hydrous oxides and organic matter in the soil, inhibiting bioavailability (Fordyce et al. 1998).

In addition to the soil geochemistry, other factors such as methods of plant uptake also influence the migration of elements from the environment through the human food chain. Muramatsu et al. (1995) working in Japan demonstrated that the soil-to-plant transfer factor for iodine in rice is very poor compared to green leafy vegetables, and that iodine in soil can be volatilized as organic/methyl iodine as a result of rice cultivation. The atmosphere is an important source of iodine in plants and atmospheric adsorption rather than soil–root uptake may contribute to rice iodine concentrations. As a consequence, concentrations of iodine in rice are often very low compared to the soils and to other crops.

Due to a combination of the soil geochemistry and poor soil-to-plant transfer ratios, total iodine concentrations in Sri Lankan rice samples were very low (< 40 µg/g in all but two samples). Therefore, despite forming the bulk of food intake, rice does not constitute a significant source of iodine in the Sri Lankan diet (Fordyce et al. 2000). An understanding of the biogeochemical factors controlling iodine uptake into the food chain is essential if effective environmental remediation strategies are to be developed.

The Future

These studies emphasize the need for better information about the distribution and behavior of iodine in the environment. This knowledge would not only increase our understanding of current environmental intervention schemes, such as the iodination of irrigation waters (DeLong et al. 1997), but could lead to the development of agricultural practices that make more efficient use of iodine already present in the environment to provide additional methods in the fight against IDD. Such schemes require collaboration between geoscientists, agricultural and veterinary specialists, and health experts. The British Geological Survey (BGS) in partnership with health and agricultural scientists recently commenced a three-year project to address some of the environmental controls on IDD and to make resources relating to iodine behavior in the environment more available to researchers in this field. Information will be disseminated from the project website at <http://www.bgs.ac.uk/dfid-kar-geoscience/idd>. This website links to, and will complement, more medically oriented information sources such as the ICCIDD.

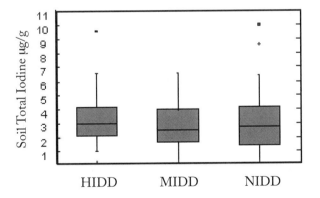

Figure 9.3: Box and whisker plots of the 10th, 25th, 50th, 75th, and 90th percentiles of soil total iodine concentrations classified by goiter prevalence.
NIDD = No/low < 10% goiter prevalence; MIDD = Moderate 10 – 25% goiter prevalence; and HIDD = High > 25% goiter prevalence.

Acknowledgments

The work presented in this chapter is sponsored by the U.K. Department for International Development (DFID) projects R6227 and R7411.

The authors gratefully acknowledge the assistance of Dr R Fuge (University College of Wales, Aberystwyth, U.K.), Dr R Benzing (Environmental Analysis Section, Imperial College Centre for Environmental Technology, Silwood Park, U.K.) and Dr Mark Cave (BGS) for analysis; Dr Neil Breward (BGS) and Dr Alex Stewart (Wirral Health Authority, U.K.) for comments on the text and Dr ABC Amarasinghe, Dept of Biochemistry, University of Peradeniya, Sri Lanka for advice on the prevalence of IDD in Sri Lanka.

Published with permission from the Director of the British Geological Survey (NERC).

References

Alicke, B., Hebestreit, K., Stutz, J. and Platt, U. (1999) Iodine oxide in the marine boundary layer. Nature, 397, 6720, p.572–573.

DeLong, G.R., Leslie, P. W., Wang, S. H., Jiang, X. M., Zhang, M. L., Rakeman, M. A., Jiang, J. Y., Ma, T. and Cao, X. Y. (1997). Effect on infant mortality of iodination of irrigation water in a severely iodine-deficient area of China. Lancet, 350, 9080, p.771–773.

Fernando, M.A., Balasuriya S., Herath K.B., Katugampola, S. (1987). Endemic goiter in Sri Lanka. In: Dissanayake, C.B., Gunatilaka, L., editors. Some Aspects of the Environment of Sri Lanka. Sri Lanka. Association for the Advancement of Science, Colombo, p.46–64.

Fordyce, F.M., Johnson, C.C., Navaratne, U.R.B., Appleton, J.D. and Dissanayake, C.B. (1998). Studies of Selenium Geochemistry and Distribution in Relation to Iodine Deficiency Disorders in Sri Lanka. British Geological Survey Overseas Geology Series Technical Report WC/98/23.

Fordyce, F.M., Johnson, C.C., Navaratne, U.R B., Appleton, J.D. and Dissanayake C.B. (2000). Selenium and iodine in soil, rice and drinking water in relation to endemic goiter in Sri Lanka. The Science of the Total Environment, v.263, p.1–3, 127–142.

Fuge R. (1996). Geochemistry of iodine in relation to iodine deficiency diseases. In: Appleton, J.D., Fuge R., McCall, G.J.H., editors. Environmental Geochemistry and Health, Geological Society Special Publication. v.113, p.201–212.

Fuge, R., Andrews, M.J. and Johnson, C.C. (1986). Chlorine and iodine, potential pathfinder elements in exploration geochemistry. Applied Geochemistry v.1, p.111–116.

Fuge, R., Johnson, C.C., Phillips, W.J. (1978). An automated method for the determination of iodine in geochemical samples. Chemical Geology v.23, p.255–265.

Hetzel B. S. (1989) The Story of Iodine Deficiency. Oxford University Press, Oxford.

Hetzel, B.S. and Maberly, G.F. (1986). Iodine. In: Mertz, W., editor. Trace Elements in Human and Animal Nutrition. Academic Press, London, p.139–197.

Mahadeva, K., Senthe Shanmuganathan, S. (1967). The problem of goiter in Ceylon. British Journal of Nutrition v.21, p.341–352.

Muramatsu, Y, Yoshida, S and Ban–nai, T. (1995). Tracer experiments on the behaviour of radioiodine in the soil-plant-atmosphere system. Journal of Radioanalytical and Nuclear Chemistry, v.194, p.2, 303–310.

Schmitz, K, and Aumann, D.C. (1995). A study on the association of two iodine isotopes, of natural I–127 and of the fission product I–129, with soil components using a sequential extraction procedure. Journal of Radioanalytical and Nuclear Chemistry, Articles, 198, 1, p.229–236.

Stanbury, J. and Hetzel, B. S. (1980). Endemic Goiter and Endemic Cretinism Iodine Nutrition in Health and Disease. Wiley, Chichester.

Tuttle, R.M. and Becker, D.V. (2000). The Chernobyl accident and its consequences: update at the millennium. Seminars in Nuclear Medicine. 30, 2, p.133–140

Stewart, A.G. and Pharoah, P.O.D. (1996). Clinical and epidemiological correlates of iodine deficiency disorders. In: Appleton, J.D., Fuge R., McCall, G.J.H., editors. Environmental Geochemistry and Health, Geological Society Special Publication. v.113, p.201–212.

World Health Organization (1993). Micronutrient Deficiency Information System: Global Prevalence of Iodine Deficiency Disorders. WHO/UNICEF/ICCIDD, Geneva

World Health Organization. (1996) Trace Elements in Human Nutrition and Health. World Health Organization, Geneva.

10

Mercury, a Toxic Metal, and Dental Amalgam Removal

U. Lindh
Centre for Metal Biology in Uppsala, Sweden
Dept Oncology, Radiology and Clinical Immunology, Rudbeck Laboratory
Uppsala, Sweden

S-O. Grönquist, B. Carlmark, and A. Lindvall
Centre for Metal Biology in Uppsala, Rudbeck Laboratory, Uppsala, Sweden

Introduction

Mercury is an element with unique physical and chemical properties whose deleterious effects on various organ systems have been known for centuries. The metal (Hg^0) mercury is the only element liquid at ambient temperatures and has an extremely high vapor pressure. Natural degassing of the earth's crust by volcanoes and emissions from soils and waters are estimated to contribute on the order of 2700 to 30,000 tons per year (Nriagu 1989, Lindqvist 1991). A second source of mercury is anthropogenic from burning of coal or petroleum. The total input into the atmosphere may be up to 150,000 tons per year, with natural emissions accounting for the major input (Berlin 1986). However, estimations of contributions from different sources vary.

Aristotle wrote about mercury as liquid silver (hydrargyrum) with the metallic mercury extracted in ancient times, as today, from the sulphide mineral cinnabar (HgS). Although technical developments have brought about more sophisticated methods of distilling mercury, all processes create mercury vapor, which is a potential hazard. Mercury mines pose environmental concern, due to mine tailings and waste rock contributing mercury-enriched sediment to watersheds (Rytuba 2000) such as in the California Coast Ranges (Rytuba 2000), the Idria mine in Slovenia (Hines et al. 2000), in Slovakia (Svoboda et al. 2000), and, perhaps most conspicuously, the mine tailings in Aznacollar, Spain, that caused a recent accident (Grimalt et al. 1999). Any industrial sites that utilize mercury during production may also produce contamination of the environment (Sunderland and Chmura 2000). The possible sources of mercury exposure are presented in Table 10.1.

Table 10.1: Sources of mercury exposure by category.

Energy Related	*Population Based*	*Industrial*
Wood	Hospitals	Chlor–alkali facilities
Coal	Dental uses	Agricultural application
Refined petroleum products	Pharmaceuticals	Pulp and paper plants
	Paint application	Base-metal smelting
	Electrical goods	Gold mining
	Municipal waste disposal	Additional industrial

Amalgamation with mercury has been used as a method for beneficiation of gold and silver since Roman times. The total global release of mercury into the environment from these activities before 1930 was estimated as over 260,000 tons. Thereafter, with the introduction of cyanidation processing technology, the emissions declined (Lacerda and Solomons 1998). However, small-scale artisanal gold mining continues and is a serious hazard to largely unskilled persons in rural areas over the world. Most attention has probably been paid to the practice in the Brazilian Amazon (Malm 1998, Santos et al. 2000) where not only has the environmental impact been studied but also consequences for the general population in the affected areas (Brabo et al. 2000, Lechler et al. 2000). One study worth commenting on here is that of de Kom et al. (1998) in Suriname. They showed that the workers had high urinary mercury excretion, but the control group, a matching group of Maroons living in a non-gold mining area and not actually exposed to mercury, had significantly higher blood mercury concentrations, and attributed these differences to a higher consumption of mercury-contaminated fish by the control group. Similar problems have been identified in Tanzania (Ikingura and Akagi 2000), in Zimbabwe (Straaten 2000), and in the Philippines (Akagi et al. 2000).

An important form of exposure, be it anthropogenic or natural, is through mercury vapor. Inhaled mercury vapor is readily absorbed in the lungs to about 80% (Hursh et al. 1976), transferred to the blood, and transported as dissolved Hg^0 in the bloodstream for some time. In the uncharged state, mercury is lipid soluble and easily enters the brain through the blood-brain barrier and becomes oxidized to Hg^{2+} by the enzyme catalase. A corresponding ionization takes place in the red blood cells after passing the membrane. This ionized form of mercury is extremely toxic due to its affinity for the sulphur-containing groups such as SH-groups in biomolecules. The signs and symptoms of acute and chronic exposure to mercury vapor are compiled in Table 10.2 (Drasch 1994).

The mad hatter disease, attributed to mercury vapor exposure long before Alice in Wonderland, has recently been revisited (Dumont 1989, Wedeen 1989). Exposure of these workers was similar to that of

dental professionals who may inhale vapor as a result of working with dental amalgams as well as from their own amalgam fillings. A series of studies of urinary excretion, which is regarded as the appropriate measure of inorganic exposure, showed that such exposure is readily detected, with the expected consequences (Rojas et al. 1998, Sallsten and Barrregard 1997, Shuurs 1999).

Animal studies confirmed that exposure to mercury from dental amalgam severely affects the immune systems in rodents (Hultman et al. 1994, 1998) and in humans occupationally exposed to mercury vapor (Ngim et al. 1992, Park et al. 2000) and amalgam (Osborne and Albino 1999). Further, Echeverria et al. (1998) presented convincing evidence that adverse behavioral effects such as impairment of memory, affect on mood, as well as motor affection and adverse influence on cognitive parameters, are associated with low Hg^0 exposures within the range of that received by the general population with amalgam fillings.

The debate over using mercury as the main component in dental amalgams has been going on for more than 100 years, commencing immediately with the modern introduction in the U.S. of the material in 1833 by the French brothers Crawcour (Ring 1993). American medical dentists were against the use of mercury-containing fillings not as much for the potential mercury poisoning but more because they regarded the Crawcour brothers as charlatans (Ring 1993). In 1895, however, G. V. Black presented the "perfect" amalgam in the sense that he introduced a balanced and standardized mixture (Travers 1994) and the German stan-

Table 10.2: Symptoms of mercury (vapor) toxicity.

Acute (1 mg/m3)	*Chronic*	
Chest pains	Violent muscular spasms	
Shortness of breath	Pricking or burning sensations	
Cough	Trembling handwriting	
Coughing blood	Irritability	Excitability
Impairment of	Insomnia	Hallucinations
pulmonary function	Shyness	Nerve pain
Pneumonitis	Drowsiness	Delirium
	Fine tremor	Loss of taste
	Loss of hearing	Loss of memory
	Loss of smell	Death

dard, proclaimed in the manual *Das Füllen der Zähne mit Amalgam* (Witzel 1899), resulted more or less in a cessation of resistance. The unease continued (Roussy 1891, Tuthill 1899) and during the 1920s, the German chemist Alfred Stock warned about the danger with mercury vapor and amalgam (Stock 1926, 1939). By the late 1970s, especially intense objections to the use of mercury-containing amalgams were raised in Scandinavia, the U.S., Germany, and other countries.

During the last twenty years an increasing number of patients have sought dental and/or medical care for problems possibly associated with amalgam, often discovered by a relationship between odontological treatment and occurrence or increase of their symptoms (Table 10.3). "Metal syndrome" is a collective term describing these effects in such patients having no definitive diagnosis in spite of thorough examination and laboratory tests (Alroth-Westerlund 1985). Other causes of disease in these patients have been excluded. A recent study indicated that mercury release from dental amalgams was substantially greater than previously estimated (Brune 1985).

Selenium, atomic number 34, and residing in the same group as sulphur, is an essential trace element. It is required for proper nutrition and the recommended daily allowance has recently been suggested as 55–70 µg/day (Dwyer 2000). Selenium is known to protect against mercury toxicity (Lindh et al. 1996, Parizek and Ostadolova 1967), although the detailed mechanism is still to be unveiled. However, Gailer et al. (2000) recently suggested a structural background of the selenium mercury interaction. To better understand the

interaction, Frisk et al. (2001a, b) undertook a detailed study using a well-established human cell model.

Mercury may also exert toxicity through generation of reactive oxygen species such as hydrogen peroxide (H_2O_2). One important scavenger for removing H_2O_2 is glutathione peroxidase (GSH-Px), an enzyme being selenium-dependent. Obviously, there is a limit to the activity of GSH-Px and when it is reached damage will occur. Additionally, mercury action on GSH-Px, which contains thiol groups, will decrease activity. If selenium is not enough for the supply of GSH-Px synthesis due to interaction with mercury, even more damage would be expected. Thus, there is good reason for monitoring not only mercury but also selenium concentrations as well as the activity of GSH-Px.

The objective of the present work was to investigate how mercury released from amalgams during removal is transferred to the blood and the possible consequences. Amalgam removal requires drilling and results in the evolution of mercury vapor. We designed our experiments to employ, or not employ, additional protection for the patient and measured mercury and selenium concentrations as well as GSH-Px activity.

Materials and Methods

Five women 37–40 years of age, all showing symptoms of metal syndrome as described by Alroth-Westerlund (1985), and listed in Table 10.3, for over 6 months were selected for this study. The median number of amalgam fillings in the group was 12. An informed consent was obtained and tests of hypersensitivity against odontological materials, chiefly metals,

Table 10.3: Signs and symptoms appearing with the metal syndrome (Alroth-Westerlund et al. 1985)

General		Neurologic	Psychiatric	Oral
Malaise	Allergy-like:	Burning pains	Anxiety	Toothache
Weight loss	respiratory	Prickly sensations	Depressed states	"Burning mouth"
Muscle pain	gastrointestinal	Loss of superficial sense	Phobia	Lingua geographica
Muscle atrophy	musculoskeletal	Spontaneous fibrillations	Impaired concentration	Discolored teeth
Headache	cardiovascular	Muscle spasms	Impaired memory	Discolored gums
Intolerance to:	skin lesions	Visual disturbances		"White lesions"
aromatic foods				Gingivitis
chemicals				Paradontitis
odors				Hyper- or hyposalivation

were negative. The patients' eating habits confirmed they ate little fish.

Two fillings, estimated weight 2–3 g each, in the lower jaw of each patient were removed. The first was removed using extra protection while the second was removed three weeks later without protection. "Without protection" is the normal dental procedure employing a high-speed drill with water spray and ordinary suction. Extra protection was achieved by adding a rubber dam around the tooth and providing an inhalation mask with mercury filter for the patient as well as employing increased suction.

Blood from a cubital vein in the crook of the arm was sampled one hour before treatment, during drilling, one hour after the amalgam filling was removed, 24 hours after, and one week after the removal. The concentration of selenium and mercury in the plasma as well as the activity of glutathione peroxidase (GSH-Px) in red cells were measured. Trace elements were measured by X-ray fluorescence (Sky-Peck and Joseph 1981) and the activity of GSH-Px was estimated us-

ing the procedures described by Paglia and Valentine (1967).

The magnitude of the patients' symptoms was subjectively estimated by the treating physician and dentist as discussed in Alroth-Westerlund (1985).

Results

General symptoms such as nettle rash and muscle pain, despite absence of hypersensitivity reactions to metals, appeared within 1 to 8 hours post amalgam removal in most of the patients. Symptoms such as debilitating fatigue and nausea were aggravated and an influenza-like condition was experienced by two patients. There were differences in symptom intensity between these patients with or without protection. The magnitude increased when unprotected within a few hours after treatment and continued for three to four days.

The measurements of selenium and mercury in the blood plasma between unprotected and protected removal of amalgam is presented in Figure 10.1. Note

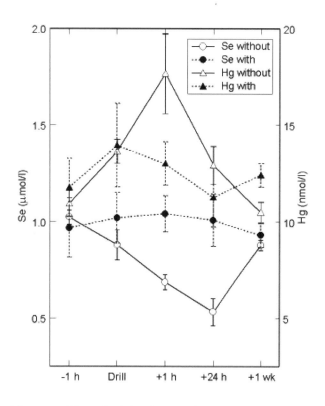

Figure 10.1: Plasma levels (median values) of Se and Hg before, during, and after removal of amalgam, with or without protection.

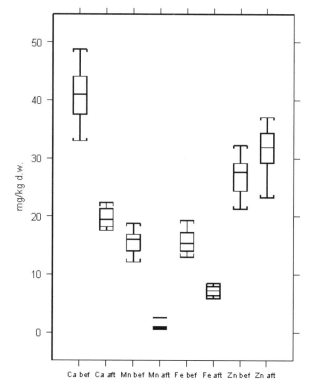

Figure 10.2: GSH-Px activity (median value) before, during, and after removal of amalgam with or without protection.

the 1000 x difference in the concentration of the elements. A healthy control group established for a broader study of the metal syndrome with essentially the same number of amalgam restorations had a selenium range of 0.6–1.7 mmol/l Se and 5.5–24 nmol/l Hg in their blood plasma.

The Hg differences are not surprising considering the rapid evolution of Hg vapor and the ease of uptake. The peak uptake (median 17.6 nmol/l) occurred one hour after drilling and returned to the starting value after one week. Extrapolating from these data, we estimate the half life of plasma mercury to 42 days in accordance with the number suggested by Kremers et al. (1999), but higher than that of Halbach et al. (1998). Our peak amount was also greater than that determined in the experiments of Halbach et al.

Assessment of the GSH-Px activity with and without protection is displayed in Figure 10.2. The GSH-Px activity reached it lowest point (median 259 μkat/l) at the same time as selenium and was delayed with respect to the mercury peak by about 24 hours in the case of no protection. Notable differences in activity between protection and no protection occur at 24 hours and 1 week. Whereas the GSH-Px activity essentially parallels the selenium concentration change, it does not return to the starting value. The corresponding activities in the healthy control group ranged from 290 to 425 μkat/l.

Discussion

Only one other study has investigated Se and Hg levels with respect to amalgam removal. Our results, a decrease in the selenium concentration in the plasma of about one day after the mercury peak, agree with those of Molin et al. (1990). This result is intriguing. Perhaps a chemical complex between selenium and mercury is formed and then disappears from the blood. However, the molar ratio between selenium and mercury is about 1000, and a one-to-one molar ratio complex formation would correspond to a reduction of selenium of about 0.9%, whereas the observed reduction was of the order of 40%. Another interpretation would be that selenium is withdrawn from the plasma pool and redistributed to the kidneys and the liver because at those sites mercury can be detoxified. There is an observed increase of selenium and mercury in liver and

kidneys when experimental animals were exposed to mercury from amalgam fillings (Hultman et al. 1998).

Mercuric mercury (Hg^{2+}) binds avidly to –SH groups and other sulphur-containing groups and can therefore affect all proteins containing thiol groups, which includes enzymes. A single ion of mercury may completely inhibit the function of the molecule. Additionally, it is known that Hg^{2+} can accelerate Fe^{2+}-induced lipid peroxidation, a process comparable to becoming rancid, and can damage cell membranes. Low concentrations of Hg^{2+} may cause production of hydrogen peroxide or, possibly, the superoxide anion (Contrino et al. 1988), an extremely reactive species that may occasion severe damage to cells as well as being implicated in carcinogenesis (Liu et al. 2001).

The interaction between mercury and selenium has been interpreted as a protective effect. The hypotheses of how this protection could be achieved in vivo involves reduced glutathione (Naganuma and Imura 1983), an abundant thiol-containing tripeptide and major intracellular reductant. As such, it maintains the sulphydryl groups of enzymes and other molecules in a reduced state.

It is interesting that the decrease in the activity of GSH-Px (Fig. 10.2) in the red blood cells is delayed in relation to the peaking of the mercury concentration (Fig. 10.1). To affect the activity of GSH-Px, mercury has to enter the red cells and be oxidized by catalase to Hg^{2+}. Part of the delay may be accounted for by this fact. However, other factors have to be involved. GSH-Px is, though, not the only target for mercury toxicity in the red cells.

The lowest concentration of selenium in the blood plasma occurs at the same time as the activity of GSH-Px is at its lowest point. Since GSH-Px is selenium dependent, this might cause a depletion of the available selenium pool inducing restricted activity of the enzyme.

The pattern of damage caused by mercury as well as the pathways of selenium protection are likely to be very complicated. This stands out ever more clearly when the affinity of mercury ions for sulphur-containing groups is taken into account. The affinity for such groups varies among metals but mercury is in a class by itself with the highest affinity of all metals. A consequence of this is that virtually all enzymes containing the amino acids methionine and cysteine are open for attack and a complex chain of events is to be expected. A detailed account of the protective mechanism will be presented elsewhere.

Conclusions

Exposure to mercury in all probabilities creates a vicious circle. Direct attachment to protein molecules and indirect effects interfere with proper functioning of cells.

When the patients were protected, the changes of plasma concentrations of trace elements and activity of GSH-Px in red cells displayed approximately the same pattern as without protection but the magnitude differed. This clearly shows that exposure to mercury during removal of amalgams can be significantly reduced but not totally eliminated using protective measures. Figures 10.1 and 10.2 indicate that it was not possible to completely avoid side effects of the treatment. Other measures have to be designed to increase the efficacy of protection.

We conclude that exposure to mercury seriously influences the symptom intensity in patients suffering from the metal syndrome and plan to continue studies on exposures to metals from amalgams and the symptoms of this disease.

References

Akagi H., Castillo E.S., Cortes-Maramba N., Francisco-Rivera A.T., Timbang T.D. Health assessment for mercury exposure among school children residing near a gold processing and refining plant in Apokon, Tagum, Davao del Norte, Philippines. Sci Total Environ 2000; v.259, p.31–43.

Alroth-Westerlund B., Carlmark B., Grönquist S-O., Johansson E., Lindh U., Theorell H., de Vahl K. Altered distribution patterns of macro- and trace elements in human tissues of patients with decreased levels of blood selenium. Nutr Res 1985;Suppl I, p.442–50.

Berlin M. Mercury. In: Handbook on the Toxicology of Metals. 2nd ed. Eds: Friberg L, Nordberg G.F., Vouk V. Elsevier, Amsterdam 1986, p. 387–435.

Brabo E., Santos E.C., Jesus I.M., Mascarenhas A.F.S., Faial K.F. Mercury contamination of fish and exposures of an indigeous community in Pará State, Brazil. Environ Res A 2000; v.84, p.197–203.

Brune D., Evje D.M. Man's mercury loading from dental amalgam. Sci Total Environ 1985; v.44, p.51–3.

Contrino J., Marucha P., Ribaudo R., Ference R., Bigazzi P.E., Kreutzer D.L. Effects of mercury on human polymorpho-nuclear leukocyte function in vitro. Am J Pathol 1988; v.132, p.110–18.

de Kom J.F.M., Voet G.B., Wolff F.A. Mercury exposure of maroon workers in the small scale gold mining in Suriname. Environ Res A 1998; v.77, p.91–97.

Drasch G.A. Mercury. In: Handbook on Metals in Clinical and Analytical Chemistry. Eds: Seiler H.G., Sigel A., Sigel H. Marcel Dekker, Inc, New York 1994, p: 479–93.

Dumont M.P. Psychotoxicology: the return of the mad hatter. Soc Sci Med 1989; v.29, p.1077–82.

Dwyer J. Old wine in new bottles? The RDA and the DRI. Nutrition 2000; v.16, p.488–92.

Echeverria D., Aposhian H.V., Woods J.S., Heyer N.J., Aposhian M.M., Bittner Jr A.C., Mahurin R.K., Cianciola M. Neurobehavioral effects from exposure to dental amalgam Hg⁰: new distinctions between recent exposure and Hg body burden. FASEB J 1998; v.12, p.971–80.

Frisk P., Yaqob A., Nilsson K., Carlsson J., Lindh U. Influence of selenium on mercuric chloride cellular uptake and toxicity indicating protection. Studies on cultured K–562 cells. Biol Trace Elem Res 2001a, in press.

Frisk P., Yaqob A., Nilsson K., Lindh U. Selenite or selenomethionine interaction with methylmercury on uptake and toxicity showing a weak selenite protection. Studies on cultured K–562 cells. Biol Trace Elem Res 2001b, in press.

Gailer J., George G.N., Pickering I., Madden S., Prince R.C., Yu E.Y., Denton B., Younis S., Aposhian H.V. Chem Res Toxicol 2000; v.13, p.1135–42.

Grimalt J.O., Ferrer M., Macpherson E. The mine tailing accident in Aznalcollar. Sci Total Environ 1999; v.242, p.3–11.

Halbach S., Kremers L., Willruth H., Mehl A., Welzl G., Wack F.X., Hickel R., Greim H. Systemic transfer of mercury from amalgam fillings before and after cessation of emission. Environ Res 1998; v.77, p.115–23.

Hines M.E., Horvat M., Faganeli J., Bonzongo J-C.J., Barkay T., Major E.B., Scott K.J., Bailey E.A., Warwich J.J., Lyons W.B. Mercury biogeochemistry in the Idrija river, Slovenia, from above the mine in to the Gulf of Trieste. Environ Res A 2000; v.83, p.129–139.

Hultman P., Johansson U., Turley S.J., Lindh U., Enestrom S., Pollard K.M. Adverse immunological effects and autoimmunity induced by dental amalgam and alloy in mice. FASEB J 1994; v.8, p.1183–90.

Hultman P., Lindh U., Horsted-Bindslev P.. Activation of the immune system and systemic immune-complex deposits in Brown Norway rats with dental amalgam restorations. J Dent Res 1998; v.77, p.1415–25.

Hursh J.B., Cherian M.G., Clarkson T.W., Vostal J.J., Mallie R.V.. Clearance of mercury (Hg-197, Hg-203) vapor inhaled by human subjects. Arch Environ Health 1976; v.31, p.302–9.

Ikingura J.R., Akagi H. Monitoring of fish and human exposure to mercury due to gold mining in the Lake Victoria gold fields, Tanzania. Sci Total Environ 2000; v.256, p.39–57.

Kremers L., Halbach S., Willruth H., Mehl A., Welzl G., Wack F.X., Hickel R., Greim H. Effect of rubber dam on mercury exposure during amalgam removal. Eur J Oral Sci 1999; v.107, p.202–7.

Lacerda L.D., Salomons W.. Mercury from gold and silver mining: a chemical time bomb? Berlin: Springer, 1998, p.146.

Lechler P.J., Miller J.R., Lacerda L.D., Vinson D., Bonzongo J-C., Lyons W.B., Warwich J.J. Elevated mercury concentrations in soils, sediments, water, and fish of the Madeira River basin, Brazilian Amazon: a function of natural enrichments? Sci Total Environ 2000; v.260, p.87–96.

Lindh U., Danersund A., Lindvall A. Selenium protection against toxicity from cadmium and mercury studied at the cellular level. Cell Mol Biol 1996; v.42, p.39–48.

Lindqvist O., guest editor. Mercury as an environmental pollutant. Water, Air, Soil Pollut 1991, p.55.

Liu S.X., Athar M., Lippai I., Waldren C., Hei T.K.. Induction of oxyradicals by arsenic: Implication for mechanism of genotoxicity. PNAS 20001; v.98, p.1643–8.

Malm O. Gold mining as a source of mercury exposure in the Brazilian Amazon. Environ Res A 1998; v.77, p.73–78.

Molin M., Bergman M., Marklund S.L., Schutz A., Skerfving S. Mercury, selenium, and gluatathione peroxidase before and after amalgam removal in man. Acta Odontol Scand 1990; v.48, p.189–202

Naganuma A., Imura N. Mode of in vitro interaction of mercuric mercury with selenite to form high-molecular weight substance in rabbit blood. Chem-Biol Interact 1983; v.43, p.271–82.

Ngim C.H., Foo S.C., Boey W., Jeyaratnam J. Chronic neurobehavioural effects of elemental mercury in dentists. Br J Ind Med 1992; v.49, p.782–90.

Nriagu J.O. A global assessment of natural sources of atmospheric trace metals. Nature 1989; v.338, p.47–49.

Osborne J.W., Albino J.E. Psychological and medical effects of mercury intake from dental amalgam. A status report for the American Journal of Dentistry. Am J Dent 1999; v.12, p.151–6.

Paglia D.E., Valentine W.N. Studies on the quantitative and qualitative characterization of erythrocyte glutathione peroxidase. J Lab Clin Med 1967; v.70, p.158–69.

Parizek J., Ostadolova I. The protective effect of small amounts of selenite in sublimate intoxication. Experientia 1967; v.23, p.142–3.

Park S.H., Araki S., Nakata A., Kim Y.H., Park J.A., Tanigawa T., Yokoyama K., Sato H. Effects of occupational metallic mercury vapour exposure on suppressor-inducer (CD4+45RA+) T lymphocytes and CD57+CD16+ natural killer cells. Int Arch Occup Environ Health 2000; v.73, p.537–42.

Ring M.E. Dentistry. An Illustrated History. Harry N Abrahams, Inc, New York 1993.

Rojas M., Guevara H., Rincon R., Rodqriguez M., Olivet C. Occupational exposure and health effects of metallic mercury among dentists and dental assistants: a preliminary study. Valencia, Venezuela; 1998. Acta Cient Venez 2000; v.51, p.32–8.

Roussy M.L. Un cas d'empoisonnement par des amalgames de cuivres. Schweizerische Vierteljahrsschrift für Zahnhelikunde 1891; v.1, p.97–103.

Rytuba J.J. Mercury mine drainage and processes that control its environmental impact. Sci Total Environ 2000; v.260, p.57–71.

Sallsten G., Barregard L. Urinary excretion of mercury, copper and zinc in subjects exposed to mercury vapour. Biometals 1997; v.10, p.357–61.

Santos E.O., Jesus I.M., Brabo E.S., Loureiro E.D., Mascarenhas A.F.S., Weirich J., Câmara M.V., Cleary D. Mercury exposures in riverside Amazon communities in Pará, Brazil. Environ Res A 2000; v.84, p.100–107.

Schuurs A.H. Reproductive toxity of occupational mercury. A review of the literature. J Dent 1999; v.27, p.249–56.

Sky-Peck H.H., Joseph B.J. Determination of trace elements in human serum by energy-dispersive X-ray fluorescence. Clin Biochem 1981; v.70, p.126–31.

Stock A. Die chronische Quecksilber- und Amalgamvergiftung. Zahnärtzl Rundsch 1939; v.48, p.403–7.

Stock A. Die Gefährlichkeit des Quecksilberdampfes und der Amalgame. Z Angew Chemie 1926; v.39, p.984–9.

Stock A. Die Gefährlichkeit des Quecksilberdampfes. Z Angew Chemie 1926; v.39, p.461–6.

Straaten P. Mercury contamination with small-scale gold mining in Tanzania and Zimbabwe. Sci Total Environ 2000b; v.259, p.105–113.

Sunderland E.M., Chmura G.L. An inventory of historical mercury emissions in the maritime Canada: implications for present and future contamination. Sci Total Environ 2000; v.256, p.39–57.

Svoboda L., Zimmermannová K., Kalac P. Concentrations of mercury, cadmium, lead and copper in fruiting bodies of edible mushrooms in en emission area of a copper smelter and a mercury smelter. Sci Total Environ 2000; v.246, p.61–67.

Travers B., ed. World of Invention. History's Most Significant Inventions and People Behind Them. Thomson Learning, Gale Group Inc, Farmington Hills 1994: p. 194–5.

Tuthill J.Y.. Mercurial neurosis resulting from amalgam fillings. The Brooklyn Medical Journal 1899; v.12, p.725–42.

Wedeen R.P.. Were the hatters of New Jersey "mad"? Am J Ind Med 1989; v.16, p.225–33.

Witzel A. Das Füllen der Zähne mit Amalgam. Nebst einem Anhange über Die moderne Behand lung pulparkranker Zähne. Berlinische Verlagsanstalt, Berlin 1899: p. 1–65.

11

Nuclear Accumulation of Mercury in Neutrophil Granulocytes Associated with Exposure from Dental Amalgam

A. Lindvall, A. Danersund, R. Hudecek
Centre for Metal Biology, Uppsala, Sweden

U. Lindh
Centre for Metal Biology in Uppsala, Sweden
Dept Oncology, Radiology and Clinical Immunology, Rudbeck Laboratory
Uppsala, Sweden

Introduction

Corrosion and wear of dental amalgam may be associated with unexpectedly high levels of endogenous exposure to heavy metals. According to WHO, the resulting uptake of mercury exceeds that from all other sources in persons not occupationally exposed (WHO 1991) and a daily uptake level of 100 µg has been reported (Barregard et al. 1995). Due to the distribution patterns of mercury, standard blood and urine analyses give meager information on the response of the organism to this exposure.

We here present data from nuclear microscopy analysis of neutrophil granulocytes (short-lived cells in the immune system cascade) in peripheral blood. Blood samples were drawn from patients suffering from possible side effects from dental amalgam. Their symptoms resembled those of the Chronic Fatigue Syndrome (Fukuda 1994), and the onset or intensity of the symptoms were related to occasional increased exposure to dental amalgam e.g., unprotected placing/ drilling of the material. Data showed profound derangement of several cellular trace elements in the patient group and in some cases the substitution of mercury for zinc in the nuclear area. This supports the contention that systemic side effects from dental amalgam may occur.

Materials and Methods

Venous blood samples were drawn from a cohort of Caucasian patients ($n = 25$) with a chronic debilitating illness, possibly related to the exposure from dental amalgam, and cell preparations were done as previously described (Johansson 1984, Lindh 1997). The same procedure was performed on blood samples from an age- and sex-matched healthy control group ($n = 22$) with similar numbers of amalgam fillings. A freeze-dried monolayer preparation of neutrophil granulocytes from each subject was investigated by means of nuclear microscopy (Zidenberg-Cherr). Thirty cells from each subject were analyzed in a subsequent manner and means and variances of the elemental concentrations of calcium (Ca), manganese (Mn), iron (Fe), zinc (Zn), and mercury (Hg) were calculated. In addition, the intracellular distribution of zinc and mercury was investigated in a few cells from both patients and controls by means of a nuclear microscopy elemental mapping technique. Cells with mercury levels above the detection limit (0.5 μg/kg dry weight) were investigated as well as cells with no detectable mercury.

One year after the removal of dental amalgam from the patients, they were examined again, and the results were compared with previous sets of data.

Results and Discussion

Comparison of the initial results from patients and controls (Fig. 11.1) showed, in the patient group, below normal cellular levels of zinc but markedly elevated levels of calcium, manganese, and iron. The difference was significant ($p < 0.001$) for all these elements.

Comparison of pre- and post-treatment data in the patient group (Fig. 11.2) showed that the cellular concentration of calcium, manganese, iron, and zinc had changed and that the difference was significant ($p < 0.001$) for all elements. The post-treatment levels were indistinguishable from the corresponding concentrations of elements in the control group.

No mercury was detected in granulocytes from anyone in the control group. In most of the patients (20/25), however, at least 2 out of 20 cells from each case had detectable levels of this element. Some of the cells with detectable levels of mercury were subjected to further investigation with elemental mapping of zinc and mercury. This revealed that mercury had entered parts of the nucleus and that zinc had been depleted in these areas (Fig. 11.3). Moreover, nuclei in such cells were hyper-segmented with five or more segments.

Since zinc and mercury are chemically similar, it could be assumed that mercury had substituted for zinc in the nucleus, and be detrimental for the nucleus and for the survival of the cell. This observation supports the hypothesis of mercury exchange for zinc with ensuing radical-generating power actually into the DNA. Transition metal-DNA interactions are believed to be the basis for toxic effects from long-term exposure

(Andersen 2000). Also, loss of zinc may impair the function of more than 200 enzymes necessary for the proper processes of the cell nucleus, e.g., DNA- and RNA-polymerases, zinc-finger proteins, and the oncgene suppressor protein p53.

From the present study we believe that chronic low-dose mercury exposure from dental amalgam may lead to nuclear accumulation of mercury and depletion of zinc in neutrophil granulocytes. This may be the basis for further cellular pathology to develop due to the multitude of zinc-dependent processes normally taking place in the nucleus. Given a certain extent of damage over a prolonged period of time, transformation into a complex clinical condition, as in our patients, should be expected.

References

Andersen B. Transition Metal-DNA Interactions. Oligonucleotides as Model Systems for Studying Platinum(II) Anticancer Drugs, with a Focus on Structure-Activity Relationships. Thesis. Bergen: University of Bergen, 2000:180pp.

Barregard L, Sallsten G, Jarvholm B. People with high mercury uptake from their own dental amalgam fillings. Occupat Environ Med 1995; v. 52, p.124–8.

Fukuda K, Straus SE, Hickie I, Sharpe MC, Dobbins JG, Komaroff A. The chronic fatigue syndrome: a comprehensive approach to its definition and study. International Chronic Fatigue Syndrome Study Group. Ann Intern Med 1994; v.121, p.953–9.

Johansson, U. E., Gille L. Application of the nuclear microprobe to the study of elemental profiles in individual blood cells. Preparation and analysis. Nucl Instr and Meth 1984; B3, p.631–6.

Lindh U, Frisk P, Nyström J, Danersund A, Hudecek R, Lindvall A, Thunell S. Nuclear microscopy in biomedical analysis with special emphasis on clinical metal biology. Nucl Instr and Meth 1997; B130, p.406–18.

WHO (World Health Organization). Inorganic mercury. Environmental Health Criteria. International Programme on Chemical Safety. Geneva, 1991: 168 pp.

Zidenberg-Cherr S, Keen CL. Essential trace elements in antioxidant processes. In: Dreosti IE, editor. Trace Elements, Micronutrients, and Free Radicals. Clifton: Humana Press, 1991, p.107–27.

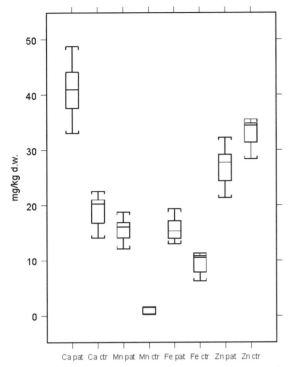

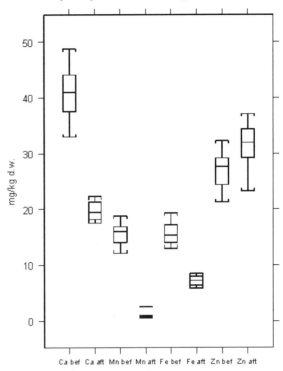

Figure 11.1: Cellular levels of calcium, manganese, iron, and zinc in neutrophil granulocytes from patients before amalgam removal (pat, n = 25) and controls (ctr, n = 22) respectively, summarized in a box-whisker diagram.

Figure 11.2: Cellular levels of calcium, manganese, iron, and zinc in neutrophil granulocytes from patients (n = 25) before (Ca bef etc.) and after amalgam removal (Ca aft etc.), summarized as box-whisker plots.

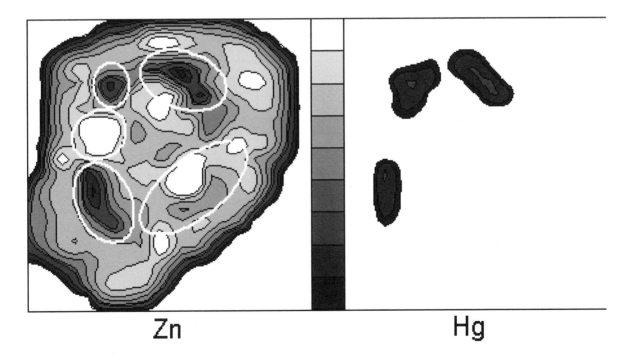

Figure 11.3: Elemental maps of zinc and ab mercury in a granulo-cyte from a patient. The shaded areas in the zinc map correspond to segments of the nucleus and are low in zinc. Three of the areas also show incorporation of mercury.

12

Cadmium Accumulation in Browse Vegetation, Alaska — Implications for Animal Health

L. P. GOUGH

U.S. Geological Survey, Anchorage, Alaska

J. G. CROCK AND W. C. DAY

U.S. Geological Survey, Denver, Colorado

Introduction

We conducted biogeochemical investigations of Cd transport and uptake by vegetation over a metamorphic and intrusive terrane in the Fortymile River watershed and Mining District, east-central Alaska. The occurrence of Cd in eolian-dominated sub-arctic soils developed over five major rock units was examined, as well as its relative bioaccumulation in willow. Although the bioaccumulation of Cd by willow (Salix spp.) has been known for some time (Gough 1991), the connection to adverse animal health, under natural (geogenic) conditions, has only recently been demonstrated (Larison et al. 2000, Mykelbust and Pederson 1999). We present Cd data for three soil horizons and the leaf and twig material of Salix glauca L. (grayleaf willow) collected at sites within defined rock units. The cycling of Cd and its bioaccumulation in willow are compared among rock types and soil horizons.

Results and Discussion

Cadmium in study area soils is derived from aeolian dust (loess) and the weathering of the primary bedrock. Plots of rare earth elements (REE, normalized to chondrite abundance's) in samples of A, B, and C soil horizon soils were similar to REE patterns in regional loess samples and did not correspond to bedrock patterns. Not surprisingly, therefore, we found essentially no difference in the concentration of Cd in soils developed over different lithologic units. In addition, the anthropogenic input (from mining) at sites we sampled was found to be minimal. Cadmium levels in soil (Fig. 12.1) are generally higher than that found in the study area rock types (~2 ppm)(Day 2000). In our acidic soils (pH 4.5–6.0), Cd should be mobile (readily leached), and should tend to form complexes with carbonates, hydroxides, and phosphates. Interestingly, Cd concentrations decrease with increasing soil depth — a trend directly proportional to soil organic matter content.

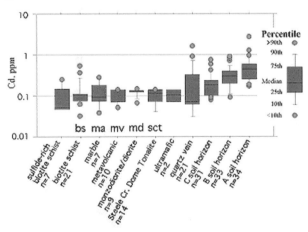

Figure 12.1: *Percentile box plots of Cd concentrations in rock units and soils of the Fortymile River watershed. Selected rock units keyed to Figures. 12.2 and 12.3 include supracrustal rocks (biotite schist, bs; marble, ma; basaltic metavolcanic, mv) and intrusive rocks (monzodiorite and diorite, md; Steele Creek Dome tonalite, sct).*

Enrichment factors (EF), a measure of the relative uptake by a plant of an element from its substrate — a sort of "bioavailability" assessment, are presented in Figure 12.2 for Cd in willow leaf material. This procedure normalizes the data, with respect to a geochemical reference element (in this case Ce), for each of the soil horizons developed over the five major rock units. For example, an EF of 5000 means that the plant is 5000 times more enriched in a given element compared to the substrate. Cadmium EF values show a marked soil-horizon difference increasing with depth. This difference reflects the greater concentration of Cd in the upper soil horizons. This trend is true for all rock units.

Cadmium concentrations in willow leaf and twig material (Fig. 12.3) were about an order of magnitude greater than those found in corresponding samples of different shrub species (Alnus spp.) and moss (Hylocomium splendens Hedw.) collected at the same sites (Gough 1999). White-tailed ptarmigan inhabiting the Colorado Mineral Belt has been shown to develop renal tubular damage that is linked to a diet of willow buds (Larison et al. 2000). Further, these authors hypothesized that Cd poisoning may be more widespread then previously suspected, among other willow-feeding herbivores (e.g., hare, beaver, and moose), in areas with high Cd in browse species. When compared to the mean of similar material collected in the Colorado Mineral Belt (2.1 ppm), our values are also elevated (mean ~1.1 ppm, Fig. 12.3). We suggest that the potential exist for a similar risk to herbivore health from Cd intake in this part of Alaska.

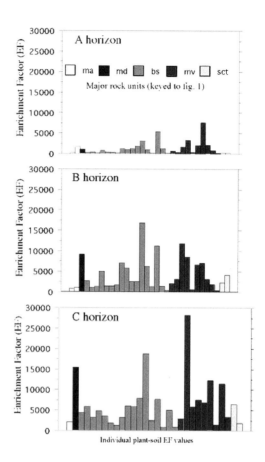

Figure 12.2: Enrichment factors for Cd in willow leaf material for A, B, and C soil horizons developed over five major rock units.

References

Day, W.C., Gamble, B.M., and Henning, M.W., and Smith, B.D. (2000) in Geologic Studies in Alaska by the U.S. Geological Survey, 1998; U.S. Geol. Survey Prof. Paper 1615.

Gough, L.P., Severson, R.C., Harms, T.F., Papp, C.S.E., and Shacklette, T.H. (1991) U.S. Geol. Survey Open-File Rept. p. 91–292.

Gough, L.P., Crock, J.G., Day, W.C., and Vohden, J. (2001) in Geologic Studies in Alaska by the U.S. Geological Survey, 1999; U.S. Geol. Survey Prof. Paper 1633 (in press).

Larison, J.R., Likens, G.E., Fitzpatrick, J.W., and Crock, J.G. (2000) Nature v. 406, p. 181–183.

Mykelbust, I. and Pedersen, H.C. (1999) Ecotoxicology 8, p.457–463.

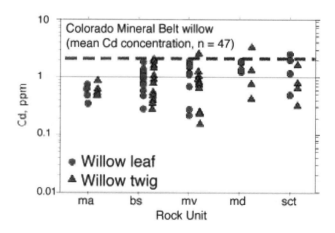

Figure 12.3: Concentrations of Cd in willow leaf and twig material growing over differing rock units (keyed to Figure 12.1) from Alaska compared to similar material from the Colorado Mineral Belt (latter from Larison et al. 2000).

13

Molybdenosis Leading to Type 2 Diabetes Mellitus
in Swedish Moose

ADRIAN FRANK

Department of Clinical Chemistry, Faculty of Veterinary Medicine,
Swedish University of Agricultural Science, Uppsala, Sweden

Introduction

The "mysterious moose disease" also called "wasting disease" is affecting moose in a strongly acidified region of southwestern Sweden (Fig. 13.1). Chemical investigations of animals from the affected region have been performed since 1988 and several articles are already published (Frank et al. 1994, Frank 1998, Frank et al. 1999, 2000a, b, c, d).

The numerous clinical signs and necropsy findings have included diarrhea, loss of appetite, emaciation, discoloration and loss of hair, apathy, osteoporosis, and neurological signs such as behavioral and locomotor disturbances (Rehbinder et al. 1991, Stéen et al. 1993). Further findings were mucosal oedema, hyperemia, hemorrhages and lesions of the mucosa in the gastrointestinal tract, hemosiderosis of the spleen and liver, dilated flabby heart, alveolar emphysema, and uni- or bilateral corneal opacity. Not all the symptoms appear simultaneously in one and the same animal. About 150–180 affected animals have been reported annually since the late 1980s.

An increase in molybdenum (Mo) and a decrease in copper and cadmium (Cu, Cd) content in organ

Figure 13.1: The region of Älvsborg (shaded) in southwestern Sweden. Source: Frank, 1998. Reprinted with permission from the publisher

79

tissues (e.g., liver) are signs of a disturbed trace element balance found in affected animals (Frank 1998) (see Fig. 13.2). To confirm the findings and to elucidate the mechanisms leading to molybdenosis and Cu deficiency, experimental studies were performed in goats. The feeding studies were performed in a controlled laboratory environment and a semi-synthetic diet was supplied (Frank et al. 2000c). Despite considerable differences in species and living conditions between goat and moose, similar changes in trace element pattern and clinical chemical parameters were observed in both species. The study shows that the etiology of the moose disease is basically molybdenosis followed by Cu deficiency, inter alia (Frank et al. 2000a,b,d).

Mo is an essential trace element that controls the metabolism of Cu in ruminants. Increased Mo concentrations relative to Cu in feed results in Cu deficiency, whereas the converse leads to an accumulation of Cu, even to Cu poisoning (e.g., in sheep). In an acidified environment, the molybdate anion is adsorbed in the soil, contrary to positively charged metals. The presence of Mo and Cu in the environment is basically dependent mainly on geochemistry, influenced by numerous physical and chemical parameters (Selinus et al. 1996, Selinus and Frank 2000). The solubility and availability of Mo to plants increases with increasing pH (e.g., after liming).

In the rumen and intestine, molybdates are transformed to thiomolybdates (TTM), which, like sulfides, bind Cu. These insoluble compounds are excreted via the feces. Excess TTM is absorbed and transported by the blood and distributed to the tissues, where available Cu is bound as a TTM complex. This inactivation of Cu-containing enzymes causes severe lesions in the animals. The various clinical signs reflect functional disturbances caused by affected Cu-containing enzymes. Cu deficiency is also known to cause glucose intolerance.

On reaching the hypothalamus, TTM causes disturbances in the endocrine balance. The hypothalamus controls appetite, the thyroid gland, synthesis of sex hormones, gluconeogenesis, and so on, resulting in weight loss, emaciation, behavioral disturbances, lethargy, and disturbances in fertility and reproduction.

Normal insulin function requires sex hormones in both males and females. Deficiency of such hormones leads to insulin resistance, a condition that may cause increased insulin production, though the efficacy of insulin is depressed and glucose concentrations may increase in the body (hyperglycemia). The long-term elevated glucose levels result in non-insulin-dependent diabetes mellitus (NIDDM), also called type 2 diabetes (Frank et al. 1999, 2000b). The increased glucose concentrations will cause glycation of proteins. These products are easily detected in diabetic humans and domestic animals, and are known to cause lesions in diabetic humans, for example in blood vessels and eyes.

In cooperation with an American group, certain reaction products [furosine, pentosidine, and carboxymethyllysine (CML)], occurring after long-term hyperglycemia, were identified and estimated in tissues of affected moose. These have been found to be higher than in non-affected individuals (Frank et al. 1999, 2000b).

Our experimental goat study revealed the principal features of this complex moose disease. The new findings have increased our understanding of the "mysterious moose disease," making it possible to interpret the clinical signs and lesions of molybdenosis and Cu deficiency in this and probably the most sensitive ruminant, the moose. From the results it could be concluded that a suggested diagnostic test for the "mysterious moose disease" ought to include determination of hepatic trace elements (Mo, Cu, Fe) and investigation of characteristic clinical chemical parameters *decreasing* in blood serum or plasma, such as ceruloplasmin, thyroxine (T_4), Mg, P_{inorg}, and *increasing* parameters, such as insulin and urea. In red blood cells, decreasing activity has been found in the enzymes Cu, Zn-SOD (superoxide dismutase), and GSH-Px (glutathione peroxidase) (Frank et al. 2000a, b, d).

This is the first time diabetes mellitus has been demonstrated with the above method in wild ruminants. It is assumed that a considerable proportion of domestic ruminants reported dead from Cu deficiency/molybdenosis (11.3 million animals in 1970, WHO) were also affected by diabetes. The method opens up new diagnostic perspectives.

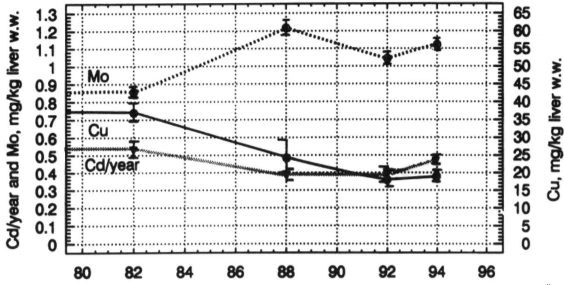

Figure 13.2: Molybdenum, copper and the average yearly cadmium uptake in the liver from moose yearlings in the southern part of Älvsborg region during 1982 through 1994. Median concentrations (median ± S.D.) are expressed in mg/kg on a wet wt. basis. Source: Frank, 1998. Reprinted with permission from the publisher.

References

Frank A. 'Mysterious' moose disease in Sweden; Similarities to copper deficiency and/or molybdenosis in cattle and sheep. Biochemical background of clinical signs and organ lesions. Sci Total Environ 1998; v.209, p.17–26.

Frank A., Anke M., Danielsson R. Experimental copper and chromium deficiency and additional molybdenum supplementation in goats. I. Feed consumption and weight development. Sci Total Environ 2000c; v.249, p.133–142.

Frank A., Danielsson R., Jones B. The mysterious disease in Swedish moose. Concentrations of trace and minor elements in liver, kidneys and ribs, haematology and clinical chemistry. Comparison with experimental molybdenosis and copper deficiency in the goat. Sci Total Environ 2000a; v.249, p.107–122.

Frank A., Danielsson R., Jones B. Experimental copper and chromium deficiency and additional molybdenum supplementation in goats. II. Concentrations of trace and minor elements in liver, kidneys and ribs, haematology and clinical chemistry. Sci Total Environ 2000d; v.249, p.143–170.

Frank A., Galgan V., Petersson L.R. Secondary copper deficiency, chromium deficiency and trace element imbalance in the moose (Alces alces L.): Effect of anthropogenic activity. Ambio 1994; v.23, p.315–317.

Frank A., Sell D.R., Danielsson R., Fogarty J.F., Monnier VM. Of moose and man: A syndrome of molybdenosis, copper deficiency, low ceruloplasmin, ataxia and diabetes. Diabetes v. 48, A284–A284. Suppl. 1 1999.

Frank A., Sell D.R., Danielsson R., Fogarty J.F., Monnier V.M. A syndrome of molybdenosis, copper deficiency, and type 2 diabetes in the moose population of south-west Sweden. Sci Total Environ 2000b; v.249, p.123–131.

Rehbinder C., Gimeno E., Belák K., et al. A bovine viral diarrhoea/mucosal disease-like syndrome in moose (Alces alces L.): investigations on the central nervous system. Vet Rec 1991; v.129, p.552–554.

Selinus O., Frank A. Medical geology. In: Möller L. (ed.). Environmental Medicine. Joint Industrial Safety Council, Stockholm. Product No. 333, 2000, p.164–183. ISBN 91–7522–634–0

Selinus O., Frank A., Galgan V. Biogeochemistry and metal biology. An integrated Swedish approach for metal related health effects. In: Appleton JD, Fuge R, McCall GJH (eds.). Environmental Geochemistry and Health in Developing Countries. Geological Society of London. Special Publication No. 113, 1996, p.81–89.

Stéen M., Diaz R., Faber E. An erosive/ulcerative alimentary disease of undetermined aetiology in Swedish moose (Alces alces L.). Rangifer 1993; v.13, p.149–156.

Part II

Anthropogenic Changes to the Geologic Environment

There are a great many ways in which human actions alter the geological environment and in doing so may cause new health risks. Industrial pollution and the disposal of chemical and domestic wastes, which alter the quality of surface and groundwater, are obvious examples. The following group of chapters deals with both general principles and specific case studies of some such hazards.

Water

Neal* surveys some scientific and policy issues surrounding surface and groundwater in the UK, including the uncertainties involved in setting environmental standards for water quality. He points out that some pollutants that entered the fluvial systems hundreds of years ago during the beginnings of the Industrial Revolution remain trapped in the sediment load of rivers, where they can act as "chemical time bombs." The surficial concentrations of harmful elements and compounds, especially heavy metals, can be remobilized during a period of rapid erosion or sediment reactivation. The potential for geological changes in streambed configuration, or in the chemical characteristics of waters, when engineering structures, such as dams and reservoirs, are built, or new areas flooded, is not fully understood, and therefore cannot yet be factored into strategies to evaluate health effects.

Point sources of potentially harmful compounds have been evaluated many times in the United States and elsewhere. People have become concerned with health effects and have sought legal assistance to impugn industrial dumping into streams. The effluent may carry heavy metals such as Hg, but a new worry is the aqueous transport of organic chemicals, as described by **Plant and Davis.** They are concerned about the growing presence of endocrine disrupters (EDCs) — suspected as important contributors to breast and prostate cancer — in soils, waters, and the human and animal food chains. They stress the importance of identifying areas of high EDC concentrations, perhaps using geochemical mapping, a theme that **Plant et al.** discuss later in this volume. Plant (2000) has recently published a book for nontechnical readers that documents her own experiences with breast cancer, and sets out her remarkable conclusions and remedies.

The fouling of water from agricultural top-dressing with fertilizers, or from leaking gasoline or household oil tanks may be relatively easily detected by flagrant algal growth or an offending smell, but there are other "silent hazards." Some, such as gasoline additives, and complex drugs like the estrogens, have a long half-life in the environment. They can go undetected because the problem area is small, localized, and goes unnoticed until obvious

** References to chapters in this volume are in **bold**.*

morphological anomalies in animals and fish are observed. Organophosphate pesticides have been regarded as safe for crops because of their fast degradation rates when exposed to surface temperatures and sunlight. Recent studies suggest that under some conditions in groundwater and soils they can persist for years, perhaps because of absorption onto soil particles. At any rate they are highly toxic and may be readily transferred to food, causing a range of diseases of the nervous and immune systems (Ragnarsdottir 2000). Such examples point to the need for constant vigilance on behalf of those responsible for maintaining acceptable domestic water quality, despite the many and diverse chemicals and the need to monitor for hazardous materials in water supplies, whether from municipal or individual wells. This is a costly process, which requires long-term commitments.

Mining

Metal concentrations in streams and groundwater around mineralized areas are commonly orders of magnitude above those in non-mineralized regions and may constitute a geogenic hazard to health (Runnells 1998). However, an additional hazard arises when the geological environment is altered artificially by mining and related activities, including exploration, refining, smelting, and metal recycling. **Grattan et al.** report on an ancient copper deposit in Jordan that was worked intermittently between 7000 and 1500 years ago. This site, which became one of the mining and smelting centers of ancient Rome, has an associated set of graves attesting to the terrible conditions endured by those early workers. The Cu content of their skeletons suggests an ultrahigh exposure to the metal. The fact that modern plants and animal tissues from this area also have relatively high contents of Cu and Pb indicates that the impact of ancient mining still influences the composition of the soils and waters in the environment today.

Robbins and Harthill, who describe the ore deposits of Minnesota and Michigan, present another long-term perspective of a mining district. This area provided the U.S. with copper, iron, and a host of other metals for over a century. The authors trace the Cu accumulations from the Precambrian to the present, characterizing them as having been localized and concentrated by a range of specialized Cu bacteria and algae. More recently, bioaccumulation of Cu has become a *cause celebre*, because of concerns about the health of indigenous people here who harvest and consume local produce.

Throughout the world's large copper provinces there are other potential health hazards to be considered, beyond those that might be repaired through restoration of the landscape. Mining creates large quantities of waste rock stored in mine dumps and tailings ponds, which may become the sources of hazardous elements or materials when exposed to the normal weathering cycle. The Cu mines of southern Arizona are excellent sites for joint geological, hydrological, biological, medical/dental studies. For example, information on mine workers' health, now required for operating the mine and associated processing areas, could be extended to include the effects of Cu concentrations on local flora, and fauna, as well as the neighboring human population. Epidemiological studies carried out in conjunction with geological and geochemical studies of weathering and transport of rock, soils, and metals from the mine area, into the surrounding water and land, could investigate the rate-limiting situations for transmission and uptake of Cu or a host of other elements. The data would also increase understanding of Cu bioaccumulation, with application to mining sites around the world.

Asbestos, which has been mined since ancient times, is now widely held to be a hazardous substance. In the twentieth century, the health effects of inhaling these fibrous materials became broadly known. **Hillerdal** reviews the evolution of the fear of the inhalation of fibrous minerals, including serpentines, amphiboles, and zeolites, and contrasts the original studies on occupational exposure to the present views of environmental or personal hazards when inhaled. Unexpected social issues arise when an area is identified with a geological hazard. **Hillerdal** shows that utilization of the zeolite-rich strata in the Cappodochio region of Turkey continues to generate bad publicity, which leads to reductions in house values, and makes more difficult the marketing of locally produced food. Many U.S. homeowners have faced similar issues on their properties.

The worldwide study of asbestosis and cancer has made asbestos a classic example of a hazardous geological material. However, chrysotile, a serpentine mineral and much the most common asbestiform material used worldwide, is actually less toxic than fibrous amphiboles. According to Gibbons (2000), the highest known rate of mesothelioma, a rare cancer directly related to asbestos exposure, is seen in South Africa, as a result of prolonged inhalation of fibrous amphiboles by black workers before preventative actions were taken by the government. Despite the existence of a wide variety of fibrous materials, many now artificially made, the precise mechanisms leading to specific health effects are not yet known (Kelse and Thompson 1989, Skinner et al. 1988, Wylie et al. 1997). To increase understanding of the detailed biomedical processes involved will require crossovers between mineralogists and materials scientists on the one hand, with cell and molecular biologists on the other.

Leaded Gasoline

It may seem strange to include here a chapter on the environmental effects of leaded gasoline. However, the review by **Mielke et al.** of the history and extent of Pb use in the U.S. and its widespread occurrence in one urban environment recalls similar situations elsewhere. The Romans depended on the lead mines in Britain, and some argue that their civilization may have foundered on the use of Pb piping for water or wine storage (Warren 2000). It is clear that ingested Pb is a possible neurotoxin with a dose response that is cumulative, that is, the longer the exposure the higher the possibilities of a harmful effect. Modern isotope geochemical analyses make it possible to distinguish the various possible environmental sources of Pb, such as discriminating between natural rock and anthropogenic sources like gas or paint. Recent environmental geochemical studies in Italy illustrate the use of Pb isotopes in establishing sources using stream sediments (Somma 1999).

Mielke et al. describe the potential harm to children of Pb deposited from gasoline in U.S. cities. Mapping the lead content of soils in playgrounds and other places where children come into contact with urban soils in New Orleans, they show differences in school performance relative to exposure levels between two schools. An obvious recommendation is for the compilation of urban geochemical maps, so that areas of concern can be identified and dealt with appropriately.

References

Gibbons, W. (2000). Amphibole asbestos in Africa and Australia: geology, health hazard and mining legacy. Journal of the Geological Society, vol. 157 (4), 851–858.

Kelse, John and Thompson, S. (1989) Regulatory and mineralogical definitions of asbestos and the impact on amphibole dust analysis. Amer. Institute of Hygiene Association Journal 50(11):613–622.

Plant, J. (2000). Your Life in Your Hands - London: Virgin, and New York: St. Martin's Press.

Ragnarsdottir, K.V. (2000). Environmental fate and toxicology of organophosphate pesticides. Journal of the Geological Society, vol. 157 (4), 859–876.

Runnells, D.D. (1998). Investigations of natural background geochemistry – scientific, regulatory and engineering issues. GSA Today, Vol. 8 (3), p10–11.

Skinner, H.C.W., Malcolm Ross and Clifford Frondel (1988) Asbestos and Other Fibrous Materials: Mineralogy, Crystal Chemistry, and Health Effects. Oxford University Press, N.Y.

Somma, R (1999) Pb isotopes and toxic metal abundances in the floodplain and stream sediments from the Volturno River Basin (Campania,Italy): natural and anthropogenic contributions. US Geological Survey, U.S. Dept. of the Interior, Reston, Va. 7 pages, 11 tables.

Warren, Christian (2000) Brush with death: a social history of lead poisoning. Johns Hopkins University Press, Baltimore, MD 362 pages

Wylie, A.G., Skinner, H.C.W., Marsh, J, Snyder, H., Garzione, C, Hodkinson, D., Winters, R. and Mossman, B.T. (1997) Mineralogical features associated with cytotoxic and proliferative effects of fibrous talc and asbestos on rodent tracheal epithelial and pleural mesothelial cells. Toxicology and Applied Pharmacology 147: 143–150.

14

Surface and Groundwater Quality and Health, with a Focus on the United Kingdom

COLIN NEAL

Centre for Ecology and Hydrology
Maclean Building, Crowmarsh Gifford, Wallingford, Oxfordshire, United Kingdom.

Introduction

Freshwater environments are of major importance to health issues in both direct (e.g., drinking water and sanitation) and indirect (e.g., industry, agriculture, and amenity/recreation) ways. However, water resources are finite, and, though renewable, demands have multiplied over the last 100 years due to escalating human populations and the growing requirements of industry and agriculture (Fig. 14.1). Hence, there are increasing global concerns over the extent of present and future good quality water resources. As Gleick (1998) emphasizes:

- Per-capita water demands are increasing, but per-capita water availability is decreasing due to population growth and economic development.

- Half the world's population lacks basic sanitation and more than a billion people lack potable drinking water; these numbers are rising.

- Incidences of some water-related diseases are rising.

- The per-capita amount of irrigated land is falling and competition for agricultural water is growing.

- Political and military tensions/conflicts over shared water resources are growing.

- A groundwater overdraft exists, the size of which is accelerating; groundwater supplies occur on every continent except Antarctica.

- Global climate change is evident, and the hydrological cycle will be seriously affected in ways that are only beginning to be understood.

The chemical composition of surface and groundwaters is influenced by a wide range of processes, some of which are outside the influence of humans while others are a direct consequence of anthropogenic pollution or changing of the environment. Starting with the range and nature of the processes involved, the changing nature of surface and groundwater quality is illustrated here, based on the evolution of the United Kingdom from a rural to an industrial and to a post-

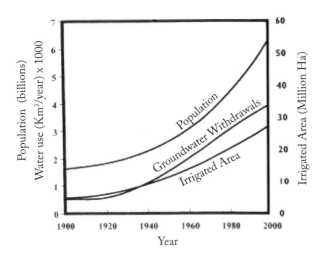

Figure 14.1: Change in the global population, water use and irrigated area over the past 100 years (data from Gleick 1998).

industrial society. The issue of what constitutes a health risk is outlined in relation to the pragmatic approaches required for environmental management.

Chemical and Hydrological Processes That Determine Surface and Ground Water Quality

Surface and groundwater exhibit a wide range of chemical compositions, and, in ecosystems uninfluenced by humans, the range of compositions can vary considerably. For example, highly concentrated $CaCl_2$ solutions (to the level of saturation, for one of the most soluble minerals known) occur at sub-zero temperatures in Antarctic lakes while brines and evaporites are associated with marine incursions and arid areas. Wide variations in pH also occur. For example, hyperalkaline spring waters of pH greater than 11 occur in ultrabasic rock terrains, whereas acidic environments associated with the oxidation of iron sulphides in soils and bedrock can produce waters of pH<3 that may be aluminum (Al) and Fe-rich. However, average compositions globally are:

- pH around neutral to slightly alkaline.

- Major ion/element concentrations greater than 1 mg/l — These comprise bicarbonate, sulphate, silica, nitrate, chloride, Ca, Mg, and Na as well as dissolved organic carbon (C).

- Minor to trace element levels in the range of 1 mg/l to 1 µg/l — K, F, P, boron (B), bromide, Sr, barium (Ba), Fe, Mn, Zn, Cu, lithium (Li), Al, and uranium (U).

- Trace element levels of less than 1µg/l — transition metals not listed above and components such as cesium (Cs), rubidium (Rb), Se, As, bismuth (Bi), and I.

- Organic components including algae and bacteria at sub µg/l levels.

Pollutants from urbanization, industry, and agriculture can change the net balance of the components. For example, mining activity can increase the concentrations of As, Fe, Cu, and others. Nonetheless, the natural geologic environment may also introduce harmful substances to groundwaters used for drinking purposes, as in the case of arsenic (As) mentioned below. Typical water chemistries for U.K. catchments affected by acid deposition and for rural, agricultural and urban/industrial rivers are presented in Table 14.1.

The chemical compositions of surface and groundwaters are linked to the bedrock sources within the catchment or aquifer, and are controlled by hydrological mixing processes (Drever 1997, Neal et al. 2000a). Wet, mist, and dry particulate deposition can be particularly important for environments where components such as sea-salt and acidic oxides from industrial emissions are supplied. Most elements in freshwaters reflect lithogenic components from "diffuse sources." They are linked to soil and bedrock type and chemical reactivity between the minerals (solids) and solution phases and the speciation in solution. Pollutant inputs come from point and diffuse sources. Point sources arise at specific locations, commonly from industry and urbanization linked to sewage works or to direct discharges, (e.g., transition metals and B). Diffuse sources are derived across a broad area and include N, P, or K fertilizers or pesticides from agriculture; and heavy metals produced by contaminated land from industrial processing and dumping.

A wide range of chemical processes determines the biogeochemical evolution of the waters within the catchment and the aquifer. These include acid-base

Table 14.1: Concentrations for water quality in four environments

Mean baseflow (left), stormflow (right) and mean (bold)
The acidic and acid sensitive catchments in mid-Wales (Afon Hafren)
The rural basins (River Tweed on the borders between England and Scotland)
The agricultural basins (River Thames in south-eastern England)
The urban/industrial complexes (River Aire, north-eastern England)

The data for the Hafren cover weekly sampling for the period 1983 to 1997 (Neal et al. 1997). Other sites cover weekly information collected from 1992 onwards (Neal and Robson 2000). Levels of micro-organic pollutants in the environment are not presented because the information is far more fragmentary (House et al. 1997, Long et al. 1998, Meharg et al. 2000a,b, and Neal et al. 2000b,c) .

		Hafren	*Tweed*	*Thames*	*Aire*
Majors (mg/l)		Acidic/Acid Sensitive	Rural	Agricultural	Urban/Industrial
	Na	4 **4** 4	16 **10** 7	52 **30** 18	164 **116** 49
	K	0.1 **0.2** 0.2	2.1 **1.4** 1.3	9.5 **5.9** 4.5	19.4 **15.2** 5.9
	Ca	0.9 **0.7** 0.7	41 **37** 20	122 **121** 116	74 **65** 36
	Mg	0.8 **0.7** 0.7	17.0 **10.2** 4.9	6.4 **5.8** 5.1	16.6 **13.8** 7.1
	NH$_4$	0.0 **0.0** 0.0	0.0 **0.0** 0.0	0.0 **0.0** 0.1	1.0 **1.3** 1.0
	Cl	7 **7** 7	26 **19** 14	65 **45** 35	179 **120** 72
	SO$_4$	4 **4** 4	16 **12** 8	110 **74** 54	212 **157** 59
	NO$_3$	1 **1** 1	1.2 **1.5** 1.5	6.1 **8.2** 9.5	9.1 **6.2** 3.7
Trace (µg/l)					
	Al	174 **311** 416	7 **24** 199	3 **6** 51	27 **17** 45
	Be	0.02 **0.05** 0.09	0.01 **0.01** 0.03	2.33 **0.02** 0.02	2.30 **0.02** 0.02
	B	5 **5** 6	46 **21** 11	333 **164** 80	575 **342** 94
	Cr	02.4 **2.4** 2.4	0.3 **0.4** 0.8	0.5 **0.4** 0.4	3.7 **3.7** 1.7
	Fe	76 **109** 127	10 **32** 167	11 **18** 75	102 **124** 204
	Mn	36 **40** 43	6 **6** 6	5 **6** 6	128 **143** 112
	Mo	0 **0** 0	0 **0** 0	4 **2** 1	42 **21** 4
	Ni	1.3 **2.0** 2.4	1.4 **1.3** 2.6	3.8 **2.7** 2.8	12.8 **8.8** 6.1
	P	0 **0** 0	33 **28** 29	1700 **818** 257	1812 **1134** 260
	Zn	14 **18** 20	6 **4** 4	8 **5** 4	32 **27** 20
	pH	5.4 **5.1** 4.5	8.3 **8.3** 8.3	8.0 **8.1** 8.0	7.5 **7.4** 7.5
	Alk (µEq/l)	0 **−21** −34	2888 **2412** 1264	4393 **4422** 4267	2583 **2252** 1351

reactions, acidity generation with mobilization of hydrolysable elements, alkalinity generation associated with biogenic CO_2 production and mineral weathering, photosynthesis/respiration, mineral solution and precipitation, partial solution/co-precipitation, complexation, oxidation and reduction, kinetic controls, ion exchange, and evaporation.

The chemical reactivity is linked to the hydrological cycle. Fluctuation in hydrological factors such as the frequency and duration of rainfall (and, in some cases, snow and mist), antecedent conditions, evaporation, and storage are commonplace. The storage term is particularly important as increased residence times allow for a greater degree of reactivity. For igneous and metamorphic rock terrains, transport of rainfall to the river occurs rapidly (minutes to days). For the more permeable sedimentary rock terrains, aquifer drainage results in a delayed rainfall response (days to months, to millennia and even longer).

For surface waters, the water quality often changes as a function of flow. Within rivers, attenuating mechanisms are linked to the volume and type of aquatic species, such as the effects of photosynthesis and respiration on pH. However, there will be a delayed response to rainfall (days to years) for lakes and reservoirs with their increased storage time. In soil waters, there will also be differences. For example, under acidic high flow conditions, these waters may contain aluminum or other metal ions, whereas groundwaters, probably richer in weathering components, would prevail under low flow conditions. For waters derived mainly from aquifers, the element concentrations vary less with

flow except when there is point source pollution that dilutes with increasing flow, or when diffuse pollutants from agriculture are leached under high flow conditions.

To illustrate the net effect of the various processes, Figure 14.2 shows the relative attenuation of the chemical elements in river water relative to the earths crust based on a study by Neal (2001a). The diagram plots the logarithm of the river enrichment factor (REF), the ratio of the concentration of each element in average surface water to the average crustal abundance. For the plot, the elements are organized in sequence from the lowest to the highest REF. For C, S, and N, the predominant species in solution are HCO_3^-, SO_4^{2-}, and NO_3^-, respectively. Note that

- The highest REFs occur for sea-salt elements (Na and Cl), soluble anions (Br and I, Se, As, and Sb) and elements related to life (C, N, and S).

- The lowest REFS occur for elements of high charge and prone to solubility controls associated with the precipitation of oxide, hydroxide (lanthanides, actinides, Si, Fe, and Mn) and layer silicate weathering products (e.g., clay minerals for Al, Si, and Fe).

- The intermediate REFS occur for components with moderate solubilities but where ion exchange (e.g. K and F) and carbonate solubility/weathering (Ca and Mg) come into play.

Water Quality in the United Kingdom: An Historical Perspective

Before 1700, Britain was a rural society with a relatively clean riverine and groundwater environment. An evolving landscape/soil/sedimentation pattern over the previous few millennia had been linked to wetter and warmer conditions, historic deforestation of much of the uplands for sheep grazing and, later, to land enclosures for farming. This produced acidic organic soil in the uplands and increased erosion in both the uplands and the lowlands.

Thereafter, Britain experienced a move from cottage to mechanized production in the Industrial Revolution, a change that eventually affected other countries through developments in agriculture, phys-

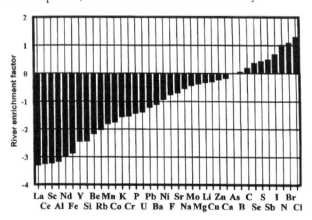

Figure 14.2: River enrichment factor (REF) for the chemical elements in global average river water (Neal 2001a).

ics, chemistry, and engineering, a fundamental restructuring of society and politics, and a changing aquatic environment (Neal 2001b). Throughout the eighteenth and much of the nineteenth century, population escalated, and major cities developed on navigable rivers. The rivers acted as a focus for trade, a source of water for drinking and industry, and a conduit for industrial and sewage wastes. Britain became the workshop of the world, and in much of the country a filthy riverine and urban environment resulted. This pattern has been repeated in some urban and industrial districts of the developing world. During the latter half of the nineteenth century and the early twentieth century, the industrial base peaked and then declined. At this time, environmental controls were progressively introduced with sewage and pollutant discharge controls and more enlightened health regulation. By the start of the twentieth century, waterborne epidemics such as cholera and typhoid became a scourge of the past.

After the 1920s, a great industrial clean-up began. This accelerated after the Second World War with the development of groups such water and purification boards and a national rivers association, expanded environmental management legislation and a reduction of industrial discharges. Over the past decade, these groups amalgamated into the Environment Agency of England and Wales and the Scottish Environment Protection Agency, major forces for environmental improvement.

Measurements of the extent of the pollution and the degree of clean-up are sparse as environmental monitoring has been undertaken only for the past 10 to 30 years. However, a numerical illustration of more recent improvements in the River Aire (Yorkshire), associated with discharge controls is shown in Figure 14.3 for the organic pollutant, dieldrin. Parallel reductions are found for many industrial chemicals as well as for trace metals such as mercury (from about 2µg/l to less than 0.1µg/l during the same period).

The broad progressive changes based on straightforward, pragmatic engineering and management measures underpinned by the developing environmental sciences, mark the actions in the U.K. as one of the worlds greatest clean-ups. However, many environmental issues remain. Groundwaters contaminated by nutrients applied as fertilizers between 1940 and 1970 are still important and contaminate lowland surface waters. Atmospheric pollution still occurs due to acidic oxides from industry (SO_x) and agriculture (NH_3), and emissions from vehicles (NO_x) are of growing importance. There also remains the legacy of the past in the form of land and river flood plain contamination by metals and micro-organic substances and drainage from derelict mines. There are the introduction of new micro-organic substances that are potentially highly harmful to the aquatic environment and the food chain. The biological quality of river and groundwater systems is of growing concern in relation to ecology and health risks from bacteria and viruses. There are issues of climate instability as changing extremes and long-term change have a major influence on how the aquatic environment functions.

Water Quality and Health

The impact of water quality on human health is highly complex. The factors involved include the environmental sources and pathways, the nature and rate of ingestion by humans, chemical concentrations, dose, and chronic and acute exposures to specific components. These factors need further consideration before approaching the issue of environmental controls in relation to health. There are three main areas to consider: inorganic components, man-made compounds or micro-organics, and biologic species that cause disease.

Inorganic Components

Most naturally occurring elements are utilized by plants and animals and some are essential for human

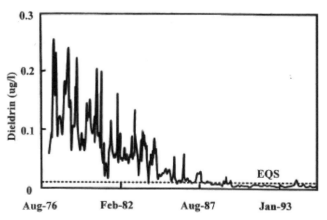

Figure 14.3: Changes in dieldrin concentrations in the River Aire: 1976 to 1994 (data from Edwards et al. 1997). EQS = environmental (water) quality standard

health and metabolism (e.g., C, S, N, P, Se, Br, I, Fe, and many of the transition metals). The issue of essential versus harmful levels is associated with the amount and the species of each element imbibed and in some cases the relationships between different elements consumed (Edmunds and Smedley 1996). In a few cases, such as for Se, F, and I, health risks have been documented resulting from deficiencies, while in other geographic areas excess ingestion is the problem. Excess imbibition of As from contaminated groundwaters in Bangladesh and China where concentration levels can exceed 600 µg/l (sixty times the new drinking water standard) has produced a major environmental problem affecting tens of millions of people. Correspondingly, in acidic, environments mobilization of Al and heavy metals within catchments and aquifers may increase Al in drinking water to 1 mg/l when the drinking water standard is 0.05mg-Al/l. In other places excess trace metals such as Fe, Hg, and Cr are the problem. Salinization of waters and influx of brines by over-abstraction of groundwaters can also lead to excess concentrations of various salts.

The impact of toxic and harmful elements on health does not relate simply to elemental composition or the bulk amount, but also to the speciation of the component. For example, Al toxicity is reduced when the element occurs complexed with humic acids or F. Correspondingly, C in the form of bicarbonate does not usually constitute a health risk, whereas cyanide (CN) clearly does even at very low concentrations, and organo-mercury compounds may be extremely toxic. In addition, the issue of biogeochemical cycling may be of central concern as toxic and other harmful components are transferred upwards through the food chain from accumulator bacterial or plant species to farm products and livestock.

Man-made Organic Compounds or Micro-organics

The sub-microgram levels per litre of micro-organics in surface and groundwaters are of growing ecological and health concern. Despite the low levels permitted by present environmental water quality standards, the wide range of industrial contaminants from point sources, and the distinct possibility that there will be both regular and intermittent discharges, present a monitoring problem and knowledge gap.

For the United Kingdom, the industrial rivers comprise a "dilute chemical soup" that is evolving over time (Neal et al. 2000a). Some substances such as DDT are banned, while others are in the process of either being tested or being banned. However, currently banned substances may still be extensively retained in flood plain sediments, and little is known about the concentrations within biological components of freshwater ecosystems. In addition, new substances are now entering river courses. At present, perhaps 100 micro-organics are being measured, although analytical scans indicate that the total number will exceed 1000. Herbicides, pesticides, and endocrine disrupters from point and diffuse sources can occur both regularly and intermittently in sewage and can, or are being, monitored. The latter include the "natural" steroid hormones (oestrone and oestradiol), "artificial" steroid hormones (ethinyl oestradiol) (**Plant and Davis**), and mimic substances (e.g., alkyl phenols and some pesticides).

Biologic Species That Cause Disease

The most serious water quality and health issues are associated with bacteria, parasites, viruses, and rotaviruses (Gleick 1998). These species proliferate in waters where there is inadequate sanitation or contact with contaminated water and from hosts that either live in water or require water. Examples are the diseases spread by insects that breed or feed near contaminated water. Globally, about six million deaths annually can be attributed to: diarrhoeal diseases — 3,300,000; intestinal helminth — 100,000; schistosomiasis — 200,000; malaria — 1,500,000; dengue fever — 20,000; trypanosomiasis — 130,000 (Gleick 1998). The disabling effects of these and other water-related diseases affect a far greater proportion of the population; the actual extent of such morbidity is unknown.

Environmental controls in Relation to Health

The United Kingdom has developed a series of environmental and health regulations over the past century. Current pollutant control strategies for river and groundwaters are based on a clear demarcation between "regulatory bodies," those responsible for maintaining/improving the water quality of the aquatic environment, and "water industry bodies" responsible for the treatment of effluents and the supplies of potable and other waters (Edwards 2001). Atmospheric emission controls, based on the concept of a critical loading to both the soil and aquatic environment, limit acidic oxide and transition metal inputs to catchments. Correspondingly, for rivers and groundwaters, targets for maximum permissible levels are set with increasing stringency in relation to ecological vitality (maintaining or improving biological diversity) in light of riverine, groundwater and estuarine fluxes, recreational activity, and water potability.

Environmental controls are largely based on water quality standards (Gardner and Zabel 1991), which specify the concentration of a substance appropriate to a specified use. These standards may be guide values, imperative values, percentile compliance, or maximum allowable concentrations in the cases where there is knowledge about the various pollutants. New substances must be registered and permission obtained for given levels of exposure. This involves setting consent levels below a "no effect level." A threshold is determined by toxicological examination employing test organisms, plus an additional safety factor (typically a factor of 10). "Trigger conditions" (the concentration at which there is a probable health impact) are required to be stipulated for each substance before registration for use. Regulations cover potable supplies, agriculture, aquatic freshwater and salt-water life, bathing, drinking water, irrigation, and livestock watering. The regulations are often complex, and even a standard table of environmental quality standards can be misleading for two reasons. First, the standards change over time. Second, the complex conditions and caveats attached to some of the water quality standards may require some discretion given to national or local authorities. The directives and circulars produced are treated as providing the optimum conditions (Gardner and Zabel 1991).

Environmental Modeling and Future Issues for Surface and Groundwater Quality Management

Management of surface and groundwater quality in relation to ecological and human health is linked to a wide range of interacting hydrological, biological and chemical interactions influenced by climate, social and industrial change there has been a move towards mathematically based simulations. To better understand how environmental systems function and change, researchers have had been benefited by the increase in computational power, new and highly sophisticated software, and the expansion of environmental databases. Within the aquatic sciences, there is now a plethora of approaches (Hauhs et al. 1996), including empirical statistically based (black box), process-orientated (lumped), process-based (deterministic/distributed), and ecosystem approaches. As the measuring capability increases, the complexity of the environmental systems becomes apparent. Generic models utilizing functional unit networks are needed for better understanding of structural uncertainty, fractal processing, and emergent properties. The simple patterns occurring at the large/catchment-outlet scale are matched by complex relationships at the local/macropore scale (Neal 1997).

Despite major advances in science and management, maintaining and improving the environment will require better understanding of the processes. Any pragmatic approach will be underpinned by environmental modeling, and the imperative to ensure "sustainable development." Two critical issues must be recognized. First, the environment is not in steady state. The contributions of climate, industry, and urban change influence hydrological, chemical, and biological environments and the extremes are probably not yet encountered. Second, dealing with and characterizing pollution and health involves many disciplines, and there are often no numeric data on which to make judgments. Environmental policy will continue to evolve (e.g., Naiman et al. 1995).

We must use innovation to solve the emerging problems, while we work to monitor the actual, and decipher the critical pollutant levels in our physical and biological environments. We need a framework for assessing the changing extremes, the relative merits of toxicological and ecotoxicological approaches, the relationship between deterministic and ecosystem approaches, and the continued development of hydrobio-geochemical models as well as how to evaluate measurements of fluxes and structural/fractal uncertainty. Environmental quality and health issues must be tackled within an integrated framework of environmental science and management, economics, politics, law, and history.

References

Drever, J.I. 1997. The geochemistry of natural waters. Prentice Hall, Upper Saddle River, NJ 07458, U.S.: 436pp.

Edmunds, W.M., and Smedley, P.L., 1996. Groundwater geochemistry and health: an overview. In Environmental geochemistry and health: eds Appleton, J.D., Fuge, R., and McCall, G.J.H.: Geol. Soc (London) special pub. 113, 91–105.

Edwards, A.M.C., 2001. River and estuary management issues in the Humber catchment. In, Land Ocean Interactions: measuring and modelling fluxes from river basins to coastal areas; eds Huntley, D., Leeks, G.J.L., and Walling, D.E. International Water Association Publications (London), SW1H 0QS, in press.

Edwards, A.M.C., Freestone, R.J., and Crockett, C.P., 1997. River management in the Humber catchment. Sci. Tot. Environ., 194/195, 235–246.

Gardiner, J., and Zabel, T., 1991. United Kingdom water quality standards arising from European Community directives - an update. Report PRS 2287-M/L WRc plc, Henley Road, Medmenham, PO Box 16, Marlow, Buckinghamshire, SL7 2HD. 27pp plus appendices.

Gleick, P.H., 1998. The Worlds Water 1998–1999. Island press, Washington DC 20009, U.S.: 308pp.

Hauhs, M., Neal, C., Hooper, R.P., and Christophersen, 1996. Summary of a workshop on ecosystem modelling: The end of an Era? Sci. Tot. Environ., 183, 1–5.

House, W.A., Leach, D.V., Long, J.L.A., Cranwell, P., Smith, C.J., Bhardwaj, C.L., Meharg, A., Ryland, G., Orr, D.O. and Wright, J., 1997. Micro-organic compounds in the Humber rivers. Sci. Total Environ., 194/195, 357–372.

Long, J.L.A., House, W.A., Parker, A. and Rae, J.E., 1998. Micro-organic compounds associated with sediments in the Humber rivers. Sci. Total Environ., 210/211, 229–254.

Meharg, A.A., Wright, J. and Osborn, D., 2000a. Chlorobenzenes in rivers draining industrial catchments. Sci. Total Environ., 251/252, 243–254.

Meharg, A.A., Wright, J. and Osborn, D., 2000b. Spatial and temporal regulation of the pesticide dieldrin within industrial catchments. Sci. Total Environ., 251/252, 255–263.

Naiman, R.J., Magnuson, J.J., McKnight, D.M., and Stanford. J.A. (editors), 1995. The Freshwater Imperative. Island Press, Washington D.C., 165pp.

Neal, C., 1997. A view of water quality from the Plynlimon watershed. Hydrology and Earth System Sciences, 1(3), 743–754.

Neal, C., 2001a. The fractionation of the elements in river waters with respect to crustal rocks: a UK perspective based on a river enrichment factor approach. Hydrology and Earth System Sciences, in press.

Neal, C., 2001b. The water quality of Eastern UK rivers: the study of a highly heterogeneous environment. In, Land Ocean Interactions: measuring and modelling fluxes from river basins to coastal areas; eds Huntley, D., Leeks, G.J.L., and Walling, D.E. International Water Association Publications (London), SW1H 0QS, in press.

Neal, C. and Robson, A.J., 2000. A summary of the river water quality data collected within the Land Ocean Interaction Study: core data for eastern UK rivers draining to the North Sea. Sci. Tot. Environ., 251/252, 585–666.

Neal, C., Wilkinson, J., Neal, M., Harrow, M., Wickham, H., Hill, L. and Morfitt, C. 1997. The hydrochemistry of the River Severn, Plynlimon, mid-Wales. Hydrol. Earth System Sci., 1(3), 583–618.

Neal, C., House, W.A., Leeks, G.J.L., Whitton, B.A., and Williams, R.J., 2000a. Conclusions to the special issue of Science of the Total Environment concerning "The water quality of UK Rivers entering the North Sea". Sci. Tot. Environ., 251/252, 557–574.

Neal, C., Jarvie, H.P., Williams, R.J., Pinder, C.V., Collett, G.D., Neal, M. and Bhardwaj, C.L., 2000b. The water quality of the Great Ouse. Sci. Total Environ., 151/152, 423–440.

Neal, C., Williams, R.J., Neal, M., Bhardwaj, C.L., Wickham, H., Harrow, M. and Hill, L.K., 2000c. The water quality of the River Thames at a rural site downstream of Oxford. Sci. Total Environ., 151/152, 441–457.

15

Breast and Prostate Cancer:
Sources and Pathways of Endocrine-disrupting Chemicals (EDCs)

JANE A. PLANT

British Geological Survey, Keyworth, Nottingham, United Kingdom

DEVRA L. DAVIS

Carnegie Mellon University, Heinz School, Pittsburgh, Pennsylvania

Cancers of the breast and prostate, which together with those of the ovaries, endometrium and testes are hormone-dependent, are among the most common forms of cancers affecting women and men respectively throughout the developed world (IARC VII 1997, Miller and Sharp 1998). The incidence of breast, prostate, and testicular cancers has risen dramatically in most European and North American countries and in Japan and Australasia since cancer registries were first compiled in the 1960s (WHO/IARC Web site). For instance, women and men born in Generation X in the U.S. and Europe today have twice the risk of developing breast and prostate cancer than their grandparents faced (Dinse et al. 1999).

Several lines of evidence indicate that environmental factors, broadly conceived, may account for some of the recent changes in patterns of hormonally dependent cancers. Although rates are about 4 times lower in Asian countries than in European ones, they are increasing most rapidly in the former (Hoel et al. 1992). People who migrate tend to develop the cancer rate of their new countries. Studies of highly exposed workers consistently find that those working with certain plastics, organic solvents, pesticides, and other toxic

chemicals tend to have higher risks of several hormonally dependent cancers (Davis and Muir 1995). Dietary factors can be involved in these patterns in two ways. Food constituents, such as dairy and animal protein products, can affect hormonal metabolism directly. In addition, foods can contain contaminants such as growth stimulating substances, pesticides, and packaging materials that can function as endocrine disrupting chemicals (EDCs).

In the past decade, a number of major national and international reports have noted the possible role of EDCs for hormone-related illnesses including breast and prostate cancer, including the Weybridge report of the European Environment Agency of 1996 (EUR 17549 1997) and the Royal Society Report on Endocrine Disrupting Chemicals (Royal Soc. 2000). This chapter presents some recent information on the sources of EDCs in the environment, outlines mechanisms by which these materials can increase the risk of hormonally dependent cancers, and discusses insights from geochemistry that may be pertinent to this work.

There have been various definitions of EDCs. The definition used here is that of the UK Environment

Agency, that "EDCs are either naturally occurring or synthetic substances that interfere with the functioning of hormone systems resulting in unnatural responses." EDCs can work through two broad pathways:

- They can directly alter hormonal metabolism by shifting the endogenous production of hormones working through hormonal receptors or other growth factors;

- They can directly damage DNA, affecting either its structure or its function.

Substances can interact with endocrine systems and cause disruption to minor functions in several ways:

- They can act like a natural hormone, bind to a receptor, and produce a similar (agonist) response by the cell.

- They can bind to a receptor and prevent a normal response, known as an antagonistic response.

- A substance can interfere with the way natural hormones and receptors are synthesized or controlled.

- EDCs or their metabolites can work through genetic pathways directly, damaging DNA to alter the cell cycle, affect repair processes, or interfere with gap-junction cellular communication.

Some of the major environmental sources of EDCs are listed in Table 15.1. Production and use of many EDCs began between the 1930s and 1950s.

Consistent with the widespread use of many of these synthetic chemicals, their residues or metabolites have been detected in many foods, rainwater, and breast milk (UKEA 2000) .

A wide variety of evidence suggests there may be a link between oestrogenic chemicals in the environment and hormone-dependent cancers. According to the Royal Society (Royal Soc. 2000), other routes of human exposure to natural estrogens that have changed in the past half century in the UK include increased consumption of dairy produce combined with changes

in dairy practice whereby pregnant cows (which produce high levels of estrogens) are in a continuous state of lactation. In addition, several studies have found that a growth factor (insulin-like growth factor-1, IGF-1) that is a small molecular-weight peptide present in milk has also been strongly implicated in the development of both breast and prostate cancer (Outwater et al. 1997). Thus, one study found that women with higher levels of IGF-1 some years earlier had a seven fold higher risk of developing breast cancer than those with lower levels (Hankinson et al.,1998) In the case of breast cancer, the main concern has been with oestrogenic chemicals, which interfere with hormonal metabolism or damage DNA through other paths. Recently, experiments in culture using a genetically modified (estrogen dependent) breast cancer cell (MCF-7) have shown that several chemicals, including bisphenol-A, PCB's, DDT, and estradiol itself, promote growth of the cells in a dose-dependent manner (Shin et al. 1999).

Many EDCs are of concern because they, or their metabolites, persist in the environment, and they are subject to large-scale transport and global circulation and distillation. They bioaccumulate up the food chain, becoming concentrated in animal products including fat meat, fish, and milk. Lindane, DDT, and other organochlorines in milk have been implicated in increased incidence of breast cancer in Israel (Westin 1993, Westin and Richter 1990). Hence, as well as being a source of natural EDCs, some foods may also be important pathways for synthetic EDCs in the environment (Westin 1993, Plant 2000). Fish kept in water containing a few parts per million of natural or synthetic estrogen develop intersex characteristics (Hoel et al. 1992). In addition, there is growing evidence of the global distribution and persistence of EDCs. For instance, Arctic animals at the top of the food chain, such as polar bears, have been found with levels of organochlorine contaminants in their fat, which would merit disposal at a hazardous waste site in the United States.

Unfortunately, the concentration and distribution of EDCs in the environment are generally unknown except at an extremely coarse scale. Even the high-risk areas have not been identified in many countries. Much more geochemical data are needed to map and

Table 15.1: Endocrine-disrupting substances (modified after the UK EA 2000 and Miller and Sharpe 1998)

Chemical	*Main Sources or Pathways*
Natural Substances	
STEROIDS	
17ß oestradiol Oestrone	Sewage-treatment works' discharges and agricultural runoff.
Synthetic Substances	
STEROIDS	
Ethinyl oestradiol (contraceptives)	Sewage-treatment works' discharges.
ALKYLPHENOLS Nonylphenol Nonylphenol- ethoxylate Octylphenol Octylphenol ethoxylate	Surfactants – in certain kinds of detergents. May enter environment via sewage. Plasticizers.
TRIAZINE HERBICIDES Atrazine Simazine	Atrazine, one of the most widely used worldwide, mainly on maize. Enters by diffuse pollution into groundwater. Non-agricultural use of both pesticides banned in many developed countries.
ORGANOPHOSPHATES Dichlorvos Dimethoate Demeton -S- methyl	Domestic, and pesticides which may enter environment through groundwater, other diffuse pollution especially through agricultural runoff, atmospheric transport, crop spraying.
ORGANIC SOLVENTS Trichloroethylen [a] Perchloroethylene [b] Benzene	Used in cleaning and degreasing operations, as lining for water pipes, in dry-cleaning, painting, and refinishing, and electronic manufacturing
ORGANOCHLORINES Endosulfan Trifluralin Permethrin	Domestic, and pesticides that may enter environment through groundwater, other diffuse pollution especially through agricultural runoff/atmospheric transport, crop spraying.
Lindane	Used as pesticide historically, including on fleeces.
DDT, aldrin, dieldrin	Still used for agriculture in some countries and may be concentrated in imported goods, but main sources in developed countries likely to be historically contaminated sites.
OTHER SUBSTANCES Polychlorinated biphenyls (PCBs)	Incineration or landfill, especially where there has been improper disposal of electrical transformers. Also a by-product of some industrial processes.
Tributyltin	Used as anti-fouling paint on large ships; enters as diffuse pollution; also a wood preservative.
Dioxins and furans	Diffuse sources including metal processing industries, medical, other waste incineration.

(a) Wartenberg et al. 2000   (b) Aschengrau et al. 1998

monitor the distribution and behavior of EDCs in the environment. In order to generate this information, new global circulation models are needed to understand the source-pathway-target relationship of natural and synthetic chemicals known or suspected to have endocrine-disrupting properties. The life cycle of such chemicals (including extraction, manufacture, transport, use, and disposal) needs to be carefully modeled and verified with observational data as a basis for a thorough estimation of the potential impacts and transformations of known EDCs. Environmental geochemistry with its extensive digital databases on ground and surface water, stream and coastal sediments, soil systems and land use patterns and research into bioavailability and speciation modeling, has much to contribute to such studies.

References:

Wartenberg D, Reyner D, Scott CS. (2000) Trichloroethylene and cancer: epidemiologic evidence. Environ Health Perspect, May;108 Suppl 2, p.161–76. Review.

Aschengrau A, Paulu C, Ozonoff D. (1998) Tetrachloroethylene-contaminated drinking water and the risk of breast cancer. Environ Health Perspect. Aug:106 Supp 4, p.947–53.

Davis DL, Muir C. (1995) Estimating avoidable causes of cancer. Environ Health Perspect, Nov; 103 Suppl 8, p.301–6.

Dinse GE, Umbach DM, Sasco A J, Hoel D G, Davis DL, (1999). Unexplained increases in cancer incidence in the United States from 1975 to 1994: possible sentient health indicators? Annu Rev Public Health, v.20, p.173–209.

EUR,1997 European Workshop on the Impact of endocrine disrupters on human health and wildlife, 2–4 December 1996, Weybridge, U.K. Report of proceedings [Report EUR 17549] (Copenhagen, Denmark: European Commission Douglas Gill X11, April 16, 1997. European Environment Agency, Kongens Nytorv 6, DK–1050 Copenhagen K, Denmark.

Hankinson SE, Willett WC, Colditz GA, Hunter DJ, Michaud DS, Deroo B, Rosner B, Speizer FE, Pollak M. (1998) Circulating concentrations of insulin-like growth factor-I and risk of breast cancer. Lancet, May 9;351(9113), p.1393–6.

Hoel DG , Davis DL, Miller AB, Sondik EJ, Swerdlow AJ. (1992) Trends in cancer   mortality in 15 industrialized countries, 1969–1986. J Natl Cancer Inst, March 4: v.84(5), p.313–20

IARC,VII, 1997 Cancer Incidence in Five Continents, 1997. Vol. V11, published by the IARC (International Agency for Research on Cancer).

Miller W.R. and Sharpe R.M., 1998 Environmental oestrogens and human reproductive cancers Endocrine –Related cancer Vol 5, p.69–96

Outwater, J.L., Nicholson, A. and Barnard, N., 1997. Dairy products and breast cancer; the IGF-1 estrogen, and bGH hypothesis. Medical Hypotheses, v.48, p.453–461.

Plant, J.A. 2000.Your Life in Your Hands. Virgin Publishing Ltd., London and St. Martin's Press, New York.

Royal Soc., 2000. The Royal Society 2000. Endocrine Disrupting Chemicals (EDCs)

Shin, K.J. and others, 1999. Development of a bioassay system for identification of Endocrine Disrupters. Proc 2nd Intern. Symp. On Advanced Env. Monitoring, Kwangju, Korea.

UKEA, 2000. The Environmental Agency, March 2000. Endocrine-disrupting substances in the environment. The Environment Agency's Strategy. The Environment Agency, Bristol, U.K.

Wartenberg, D. Reyner, D and Scott, C.S. (2000) Trichloroethylene-contaminated drinking water and the risk of breast cancer. Environ. Health Perspect., May;108 Suppl. v.2, p.161–76. Review

Westin, J.B. 1993. Carcinogens in Israeli milk: a study in regulatory failure. Inter J. Health Serv. v.23(3), p.497–517.

Westin, J.B., Richter, H. 1990. The Israeli breast cancer anomaly. Ann. NY Acad. Sci. v.609, p.264–279.

WHO/IARC web site: http://who-pcc.iarc.fr

16

A Legacy of Empires? An Exploration of
the Environmental and Medical Consequences of Metal Production
in Wadi Faynan, Jordan

JOHN GRATTAN, ALAN CONDRON, SHARON TAYLOR
Institute of Geography and Earth Sciences
The University of Wales, Aberystwyth, United Kingdom.

LOTUS ABU KARAKI
Department of Antiquities, Amman, Jordan

F. BRIAN PYATT
Department of Life Sciences, Nottingham Trent University, Nottingham, United Kingdom

DAVID D. GILBERTSON
School of Conservation Sciences, Bournemouth University, Dorset, United Kingdom

ZIAD AL SAAD
The Institute of Archaeology and Anthropology, Yarmouk University, Irbid, Jordan

Introduction

We have explored, and outline herein, the accumulation of copper in humans, plants, and animals in a remote desert area of southwest Jordan, Wadi Faynan, where mining and smelting activities began about 7000 years ago and effectively ceased 1500 years ago (Fig. 16.1). The archaeological core of the area, Khirbet Faynan, is the ruin of the Roman city of Phaino, one of the major mining and smelting centers of the Roman world. In addition, the Faynan area was one of the most important suppliers of copper to ancient Syria, Mesopotamia, and Egypt (Klein and Hauptmann 1999). Ancient industrial archaeology abounds in the form of adit and shaft mines, ore and metal processing sites, kilns, and spoil and slag heaps (Hauptmann et al. 1992, Hauptmann 2000). The industrial archaeology is closely associated with a complex and extensive irrigated system of fields, which must have been constructed and maintained to feed the workforce in this remote arid area (Barker et al. 1998, 2000). Wadi Faynan is therefore ideally suited to explore the environmental impact of metal production in the past, and its impact, if any, in the modern environment.

The Study Area

The study area is located in the hot and very arid Jordanian Desert at the mountain front at the eastern margin of the Wadi Araba, between the Dead Sea and the Gulf of Aqaba (Fig. 16.1). These environmental conditions promote the widespread deflation and redistribution of dusts, which inevitably include metalliferous materials released from eroding spoil and slag heaps and ore processing sites (Gee et al. 1997, Pyatt and Birch 1994). The geology of the region is very complex and of key importance to understanding the consequences of mining and pollution in the region. Copper and locally lead mineralization is present in several rock strata, in particular the Numaya Dolomite Limestone of the Durj Dolomite Shale Formation and the Umm' Ishrin Sandstone Formation of Middle and

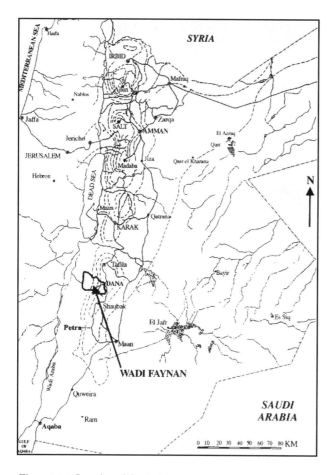

Figure 16.1: Location of the study area.

Early Cambrian age (Barjous 1992, Bender 1974, Hauptmann 2000, Rabb'a 1992).

Geochemical Records of Mining and Smelting History

Adjacent to Khirbet Faynan is a reservoir that was abandoned as a water storage facility before the fifth century BC. Since abandonment, there appears to have been a steady accumulation of fine sediments, which have now completely infilled the old reservoir basin (Barker et al. 1998). Sediments were sampled at 1 cm intervals from the cleaned face of a 2.8 m pit, excavated to bedrock, and analyzed for their lead and copper contents on the Aberystwyth ICP-MS instrument. Analytical details are presented in Perkins et al., (1995). Organic material at 230-cm depth was dated to 500 BC (Beta 110841) and the results from the sampled pit, presented in Figure 16.2, indicate that substantial localized pollution occurred in ancient times. Between 500 BC and the probable abandonment of the mines

in the sixth–sevennth century AD, copper concentration in these accumulated sediments never fell below 2000 ppm, and was frequently over 3000 ppm, coinciding with sediment depth of 163 to 237 cm. Peak productivity of the mine is coincident with the peak pollution recorded in the Greenland ice-cores and in the peat cores from northern Europe (Hong et al. 1996, Shotyk et al. 1997, West et al. 1997).

With urban and agricultural land use in close proximity to industrial processes in ancient times, not only miners and metalworkers, but agricultural workers, administrators, indeed any occupant of the district, may have been exposed to toxic concentrations of copper and other metals. Pathways by which metals could have accumulated in the ancient occupants of Wadi Faynan would have been diverse, including inhalation of metaliferrous dusts and flue gases during industrial processes and ingestion via the consumption of plants and the animals that would have ingested the local plant materials containing the metals (Leita et al. 1991, Moustakas et al. 1997, Ylaranta 1996). We explore the possible pathways operating in the past by assaying the metal (Cu and Pb) content of bone using samples of Byzantine skeletons, and for the present the results of the analyses of modern plants and animals.

Medical Impacts on Ancient Peoples

A preliminary analysis of the degree of exposure to copper has been conducted using atomic absorbtion spectroscopic analysis (AASA) of selected bones of four adult Byzantine workers, two men and two women, excavated from the extensive Byzantine cemetery in Wadi Faynan (Karaki 2000). The skeletons of all four individuals showed joint damage, which may be related to the physical nature of the work in which they were engaged. All the material analyzed (at the University of Yarmouk) was clean cortical bone. To minimize the possibility of post mortem contamination all surficial samples of bone were cleaned using silica abrasives. Samples were air dried to 105° C, finely ground, and approximately 2.5 grams of the dried bone were digested overnight in 30% conc. HNO_3, cooled, and filtered into a 25-ml flask. Analytical results are the average of 5 replicates.

The data suggest that people living in Wadi Faynan could have suffered from an overexposure to copper.

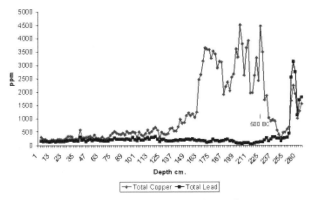

Figure 16.2: Total copper and lead extracted from sediments trapped in the Khirbet Faynam Reservoir.

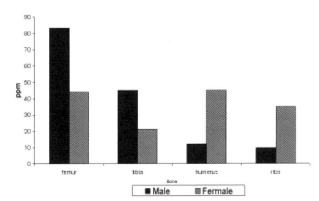

Figure 16.3: Copper concentrations in the skeletons of Byzantine metal workers.

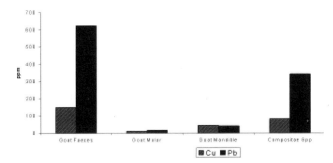

Figure 16.4: Copper and lead concentrations in modern plant and animal material.

Copper content in the bones analyzed (Fig. 16.3) was between two and eighteen times in excess of the typical figure for vertebrate bone, 4.2 mg/kg^{-1} (Scheinberg 1979), and indicates potentially toxic exposure to copper. For the males analyzed, copper appears to be selectively found in bones with the largest blood supply, those that cycle phosphate most rapidly (Skinner 2000). This pattern of distribution is less clear in the two females analyzed and the reasons for this are not yet clear. However, we know from Agatharchides, a Greek geographer who visited the mines of Arabia in the second century BC, that men and women were engaged in different activities (cited in Agricola 1556). Males were directly engaged in the mining process while females were engaged outside the mines. The copper concentrations of Figure 16.3 may indicate differences in dose related to the types of exposure. What is beyond doubt is that all the Cu concentrations determined are in excess of the typical amounts determined for copper in 'normal' human bone. There is more excavated material (Findlater et al. 1998), a total of 52 individuals. The additional samples will enable us to further explore these issues.

In particular, we plan to distinguish between the metal contents of cortical and trabecular bone to ask whether there is a differential replacement in the more rapidly turning over tissues and whether a portion of the divalent metal Ca in the mineral matter of bone is replaced by Cu. In addition, we would like to explore the hypothesis that high Cu content could also contribute to the observed incidence of joint degeneration. Moderate symptoms of copper toxicity identified today include, salivation, epigastric pain, nausea, vomiting, and diarrhea (Mason 1979). Serious problems include intravascular hemolysis, hepatic necrosis and failure, hemoglobinuria, proteinuria, hypertension, tachycardia, acute renal tubular failure, coma, and ultimately death (Williams 1982). Workers directly exposed to copper fumes and dusts have 'copper fume fever' and occasionally the ulceration of the nasal septum has also been observed (Cohen 1979). Even moderate symptoms of copper toxicity must have made life unbearable for the unfortunate souls condemned to mines, and small wonder therefore that Roman chroniclers stated that even condemned murderers were hardly likely to live more than a few days when sent to the mines of Phaino (Eusebius 1969).

The Modern Environment

The degree to which the ancient environment was affected by the pollution has been discussed in Pyatt et al. (1999 2000). However, the legacy of empire is still very evident in the modern environment (Fig. 16.4). Preliminary results indicate significant bioaccu-mulation of metals in plants and animals. The measured concentrations of copper and lead in plant and animal tissue are higher than that observed in the current topsoil. These figures may have health implications for the herbivores of the region (Bernard 1977). However, concern must also be expressed for the health of the human inhabitants, potentially the top of the food chain. While exposure of the modern inhabitants is unlikely to approach that of the ancient miners, it is beyond doubt that they live in an environment where lead, copper, and other metals continue to cycle in the environment with potential health impacts. Future research aims to determine the exposure of the modern human population to ancient pollution.

Conclusion

This chapter has explored issues of medical geology in a remote area of Jordan. It has begun to trace in more geochemical detail pollution pathways and biogeochemical cycling of copper in both the past and in the present. It is clear from the analyses that metals are actively cycling in the environment today and may be part of the human food web. The possibility of medical consequences exists and must be explored. The legacy of empires is apparent: some metals persist in the environment for thousands of years and there is the possibility that elevated metal concentrations produced in the past will impact health today and in the future. These studies in Jordan suggest there are parallel situations elsewhere in the world.

References

Agricola, G., 1556, De Re Metallica. Translated by H. Hoover and L. H. Hoover. 1950. New York, Dover Publications, p. 276.

Barjous, M.O., 1992, The Geology of the Ash Shawbak Area. Map Sheet no. 3151-III, Geological Mapping Division. Bulletin 19. Amman, Geology Directorate.

Barker, G.W., Adams, R. Creighton, O.H., Daly, P. Gilbertson, D.D. Grattan, J.P., Hunt, C.O., Mattingly, D.J., McLaren, S., Newson, P., Palmer, C., Pyatt, F.B., Reynolds, T.E.G., Smith, H., Tomber, R. and Truscott, A.J., 2000, Archaeology and desertification in Wadi Faynan:: The fourth (1999) season of the Wadi Faynan Landscape Survey: Levant, v.32, p. 27–52.

Barker, G.W., Adams, R. Creighton, O.H., Gilbertson, D.D., Grattan, J.P., Mattingly, D.J., McLaren, S., Newson, P., Reynolds, T.E.G. and Thomas, D.C., 1998, Environment and Land Use in the Wadi Faynan: Geoarchaeology and Landscape Archaeology: Levant, v.30, p. 5 – 25.

Bender, F., 1974, Geology of Jordan: Berlin, Gebr, der Borntraeger, 196p.

Bernard, S.R., 1977, Dosimetric data and metabolic models for lead: Health Physics, v.32, p. 44–46.

Cohen, S.R. 1979, Environmental and occupational exposure to copper: in Nriagu, J., ed., Copper in the Environment Pt II, Human Health: London, John Wiley, p. 1–16.

Eusebius 1969. The History of the Church from Christ to Constantine. Harmondsworth, Penguin Books, 469 p.

Findlater, G., El-Najjar, M., Al-Shiyab, A., O'Shea, M. and Easthaugh, E., 1998, The Wadi Faynan project: The south cemetery excavation, Jordan 1996: Levant, v. 30, p. 69–83.

Gee, C., Ramsey, M.H., Maskall, J. and Thornton, I., 1997, Mineralogy and weathering processes in historical smelting slags and their effect on the mobilisation of Lead: Journal of Geochemical Exploration, v.58, p. 249–257.

Hauptmann, A., 2000, Zur Fruhen matallurgie des kupfers in Fenan/Jordanien. Bochum, Deutch Bergbau Museum, 238p.

Hauptmann, A., Begemann, F., Heitkemper, E., Pernicka, E. and Schmitt-Strecker, S., 1992, Early copper produced at Feinan, Wadi Araba, Jordan: The composition of ores and copper: Archaeomaterials, v.6, p. 1–33.

Hong, S. Candelone, J-P, Patterson, C. and Boutron, C.F., 1996, History of Ancient Copper smelting pollution during Roman and Mediaeval times recorded in the Greenland ice: Science, v.272, p. 246–249.

Karaki, L.O.A., 2000, Skeletal biology of the people of Wadi Faynan: A bioarchaeological study: Unpublished MA Thesis, Jordan, Yarmouk University, 114p.

Klein, S. and Hauptmann, A., 1999, Iron Age leaded tin Bronzes from Khirbet edh-Dharih, Jordan: Journal of Archaeological Science, v.26, p. 1075–1082.

Leita, L., Enne, G., De Nobile, M., Baldini, M. and Segui, P., 1991, Heavy metal bioaccumulation in lamb and sheep bred in smelting and mining areas of SW Sardinia, Italy: Bulletin of Environmental Toxicology and Contamination, v.48, p. 887–893.

Mason, J., 1979, A conspectus of research on copper metabolism and requirements: Journal of Nurtition, v. 109, p. 1979–2006.

Moustakas, M., Ouzounidou, G., Symeonidis, L. and Karataglis, S., 1997, Field study of the effects of excess copper on wheat photosynthesis and productivity: Soil Science and Plant Nutrition, v.43, p. 531–539.

Perkins, W.T. and Pearce, N.J.G., 1995, Mineral microanalysis by laser-probe inductively coupled plasma mass spectrometry: in Potts, P.J., Bowles, J.F.W., Reed, S.J.B. and Cave, M.R., eds., Microprobe Techniques in the Earth Sciences: London, Chapman and Hall, p. 291–325.

Pyatt, F.B. and Birch, P., 1994, Atmospheric erosion of metalliferous spoil tips: Polish Journal of Environmental Studies, v.4, p. 51–53.

Pyatt, F.B., Barker, G.W., Birch, P., Gilbertson, D.D., Grattan, J.P. and Mattingly, D.J., 1999, King Solomonís miners - starvation and bioaccumulation? An environmental archaeological investigation in southern Jordan: Ecotoxicology and Environmental Safety, v.43, p. 305–308.

Pyatt, F.B., Gilmore, G., Grattan, J.P., Hunt, C.O. and McLaren, S., 2000, An Imperial legacy? An exploration of Ancient metal mining and smelting in Southern Jordan: Journal Of Archaeological Science, v.27, p. 771–778.

Rabb'a, I., 1992, The geology of the Al Quraryqira (Jabal Hamra Fadda). Map Sheet 305 II. Amman, Geology Directorate, Geological mapping Division, Bulletin 28.

Scheinberg, I.H., 1979, Human Health effects of copper: in Nriagu, J., ed., Copper in the Environment Pt II, Human Health: London, John Wiley, p. 17–39.

Shotyk. W., Cheburkin, A.K., Appleby, P.G., Fankhauser, A., and Kramers. J.D., 1997, Lead in three peat bog profiles, Jura Mountains, Switzerland. Enrichment Factors, Isotopic composition and chronology of Atmospheric Deposition: Water, Air and Soil Pollution, v. 100, p. 297–310.

Skinner. H.C.W., 2000 Minerals and Human Health. p.383–412 in Environmental Mineralogy Edited by D. J. Vaughan and R. A. Wogelius EMU Notes in Mineralogy, Eotvos University Press, Budapest, Hungary.

West, S., Charman, D.J., Grattan, J.P. and Cherburkin, A.K., 1997, Heavy metals in Holocene peats from South West England: detecting mining impacts and atmospheric pollution: Water, Air and Soil Pollution, v. 100, p. 342–353

Williams, D.M., 1982, Clinical significance of copper deficiency and toxicity in the world population: in Prasad, A.S., ed., Clinical, biochemical and nutritional aspects of trace elements: New York, A.R.Liss, p. 277–299.

Ylaranta, T., 1996, Uptake of heavy metals by plants from airborne deposition and polluted soils: Agriculture and Food Science in Finland, v.5, p. 431–477.

17

Life in a Copper Province

ELEANORA I. ROBBINS AND MICHALANN HARTHILL
U.S. Geological Survey, Reston, Virginia

Introduction

The North American Lake Superior region contains a world-class copper province nearly 300,000 km² in area (Fig. 17.1). A dozen major copper deposits and hundreds of smaller mineral accumulations are located in Michigan, Minnesota, and Ontario (Morey and Sims 1996), many of which include As, nickel (Ni), platinum (Pt), palladium (Pd), Co, Mo, and Fe, as well as Cu (Nicholson et al. 1992). Historically, the province has hosted a variety of life forms; fossil biota have been traced as far back as 2.6–2.75 Ga. Palynological and microbial research (Robbins 1985, Robbins et al. 1994) prompts speculation about possible correlations between copper and the biota that evolved there. Because the region has been subjected to continental collisions, volcanism, glaciation, rifting, weathering, sea level rise and fall, waxing and waning of lakes, soil formation, and now to human settlement and development including mining, the fossil record is discontinuous. This review of the geologic formations in the Lake Superior region from the Precambrian to the present, and their copper and biotic occurrences and associations, attempts to illuminate some of those geologic/ biologic correlations, and includes mention of modern environmental concerns.

Figure 17.1: Copper deposits and occurrences (from Cannon and McGervey 1991, Chandler et al. 1982, Nicholson et al. 1992, Sims et al. 1993).

Key: • = copper accumulations; bath. = batholith; com. = complex

Biology of Copper Utilization

Copper is one of nearly 75 chemical elements contributing to metabolic or structural functions of organisms (Dexter-Dyer et al. 1984). Bioassimilation varies and depends not only on availability from the environment, but also on the species, gender, and age of organism with specific concentrations also depending on diet, health, tissue assayed, and various synergisms with other trace elements such as Fe and Zn. Indeed, copper is an essential element and co-factor contributing to copper-associated polypeptides that provide catalytic and electron transfer functions in almost every known group of organisms alive today, from bacteria to humans. Copper proteins contribute to skin pigmentation, nerve coverings, and in mechanisms of development, maintenance, and repair of connective tissues important for well-functioning cardiovascular systems (Eisler 2000). Presently, over two dozen essential copper proteins, some with porphyrin-copper functional groups (similar to the porphyrin-iron association in hemoglobin), have been identified, each with its specific developmental or physiological function (Cowan 1998) (Table 17.1).

Adult humans contain between 1.4–2.1 mg Cu/kg of body weight, derived primarily from legumes, potatoes, nuts, seeds, and beef. The new Recommended Daily Allowance for copper for adults in the United States is 900 µg/day, with 10 mg/day the suggested upper limit (NRC 2001). Excess Cu is normally excreted. Copper toxicity, however, has been identified, such as Wilson's Disease, a rare human genetic disorder (Sternlieb 2000) and there are also non-genetic based excesses that may lead to chronic pulmonary disorders, cirrhosis, and may even result in death (Muller et al. 1998). Normally detoxification is through expulsion via cellular copper pumping (Weissman 2000) or metallothionein sequestration, a method that combines the metal with sulfhydryl groups (Fabisaik 1999). Some bacteria can precipitate excess as aqueous forms of copper on their cell walls or membranes (Robbins et al. 1994). The apparent ubiquitous requirements for copper and the ability to avoid its toxic activity suggests copper may have been available, required, and used at the origin of life, and continues to this day as essential in certain roles throughout many if not all life forms (Beck and Ling 1977).

Geologic Abundance and Mobility of Copper

Crustal and sedimentary rocks range in copper concentration from 24–45 mg/kg world-wide, although marine black shales have concentrations up to 300 mg/kg . Estimates of worldwide concentrations of copper vary from 0.2 to 30 µg/L in fresh water and 0.05 to 12 µg/L in seawater (Hem 1985). Pourbaix (1966) showed that the dissolved form of bioavailable copper is cuprous (Cu^+) in dilute waters (Cu<1 ppm) or anoxic water, cupric (Cu^{++}) in acid or neutral oxygenated waters, $HCuO_2^-$ in oxygenated alkaline water (pH >8.5), and $CuCl_2^-$, when water contains NaCl. Concentrations of total dissolved copper in Lake Superior (pH 8) ranged from 1.4 to 6.9 µg/L between 1970 and 1973 (Weiler 1978). Soils from Cambrian sandstones in the Pictured Rocks Lakeshore area, MI, have assayed 150 ppm (Schacklette 1967).

Linkages: Geological and Biological Processes

Major compilations of the geology of the Lake Superior region (Sims and Carter 1996), and distribution of copper through time (Jacobsen 1975) are available (Table 17.2). This chapter focuses instead on those rocks that couple copper with carbon or fossil materials, although 66% of stony meteorites bear native Cu Rubin (1994). The region is known also for its world-class Precambrian iron-formation deposits (LaBerge 1994), which may contain copper (Kirkham 1979). However, the important copper minerals in this region are native copper (Cu^0), chalcopyrite ($CuFeS_2$), chalcocite (Cu_2S), bornite (Cu_5S_4), domeykite (Cu_3As), and malachite ($Cu_2(CO_3)(OH)_2$).

As an organism enters the fossil record, its metal components accompany it to deposition (Honjo et al. 1982). Interpreting relationships between fossil organisms and metals enclosed within their tissues relies on a variety of microtechniques, including palynological analysis using light microscopy, scanning electron microscopy (SEM), and isotopic analysis. Black opaque copper sulfide mineral(s) were identified enmeshed in organic tissue fragments of the Proterozoic Nonesuch Shale in the region (Robbins 1983, Robbins and Traverse 1980). Similarly, Sillitoe et al. (1996) identified bacterial clusters with chalcocite that formed at the replacement fronts with chalcopyrite or pyrite in a Chilean copper deposit.

Table 17.1: Copper proteins. The table lists identified copper proteins and their functions within taxonomic categories (adapted from Cowen 1998 and Margulis and Schwartz 1998). Copper proteins require at least one copper ion or copper as a cofactor for metabolic activity. (X= Known user; ? = Possibly used; na = Information not available)

Function	Protein Names	Bacteria	Protoctista	Animalia	Fungi	Plantae
Catalysis	Amine oxidase	X	na	X	X	X
	Ammonia monooxygenase	Nitrifying bacteria	?			
	Ascorbate oxidase					Fruits and vegetables
	Ceruloplasmin			Mammals		
	Cu,Zn SOD [a]	X	na	X		
	Cytochrome c oxidase	Aerobic bacteria	X	X	X	X
	ba(3) and caa(3) [b]	X				
	Diamine oxidase	na	na	X	na	X
	Dopamine B-hydroxylase	na	na	X		
	Glactose oxidase			Mammals		
	Laccase				X	X
	Lysl oxidase			X		
	CH_4 monooxygenase	X	?			
	N_2O reductase	X				
	Nitrite reductase	X				
	Peptidylglycine hydroxylating monooxygenase			X		
	Tyrosinase			X	X	X
	Ubiquinone oxidase	X	X	X	X	X
Electron transfer	Auracyanin	Chloroflexus [c]				
	Azurin	X	na	na	na	X
	Phytocyanin family					X
	Plastocyanin family	Cyano-bacterium[d]	Green algae			X
	Rusticyanin	X				
Metallothioneins	Cu-MTN	na	na	Drosophila	X	na

(a) Cu,Zn superoxide dismutase
(b) Members of the Cytochrome C oxidase group
(c) Chloroflexus aurantiacus, green gliding primitive photosynthetic bacterium
(d) Prochlorothrix hollandica

Isotopic variations (as δCu^{65} in parts per thousand, ‰) can provide some evidence of biological fractionation. In the Lake Superior province, fractionation has been measured on native copper (−1.8 ‰), on chalcocite and on domeykite (0.0 to +1.4 ‰), and in host rock shales/siltstones/sandstones (-3.6 to +1.4 ‰) (Shields et al. 1965, Walker et al. 1958). Although multiple physical sources of copper exist, Zhu et al. (2000) interpret the isotopic variability as biological fractionation at low temperatures.

Within the large Lake Superior region, the earliest organic carbon appears in the Late Archaean (2.6–2.75 Ga) Soudan Iron-Formation of the Vermilion Range, Wawa Subprovince. The metasedimentary marine rocks were found to contain bacteria or cyanobacteria (Cloud et al. 1965). Microtechniques have yet to be applied to determine if copper mineralization is directly related to in situ organic carbon layers. In the Wabigoon Subprovince, the Ontario Atikokan area has chalcopyrite (Kirkham 1979) and spectacular domed dolomitic stromatolites created by sediment-coated cyanobacteria, possibly assisted by other bacterial species (Wilks and Nisbet 1985). Pyrite with biogenic sulfur values occurs in horizons above these stromatolites (Strauss 1986).

The Middle Proterozoic (1.0–1.6 Ga) contains fossilized algal filaments and pellet-shaped microfossils that may be the earliest organic remains in the Marquette Range of Michigan. Copper sulfide mineralization (Clark 1974, Mudrey and Kalliokowski 1993) is associated with the Kona, Negaunee, and Michigamme Formations that have organic carbon as microbial remains, stromatolites, and algal filaments (Cloud and Morrison 1980, Han and Runnegar 1992). In the Animikie Basin of Minnesota and Ontario, the Rove Formation in the Gunflint Range shows copper mineralization (Reed 1967) and microfossils of bacteria, pellet-shaped structures that may represent the oldest fecal pellets of simple multicellular animals (Robbins et al. 1985), and worms (Edhorn 1973).

The tectonic activity of the Late Proterozoic (0.6–1.0 Ga) (Cannon 1992) resulted in many copper deposits along with high biological productivity in a rift valley lake (Daniels 1982). The black shale of the Nonesuch Formation lakebeds contains petroleum, and fossil cyanobacterial sheaths, fungal hyphae-sized tissues, and aquatic algae (Robbins 1983). The variety of pellet-shaped microfossils and spore triads (Strother 1986) implies that microscopic animals may have lived and reproduced in the water column (Robbins et al. 1985). Chalcocite enmeshed within organic tissues in the shale (Robbins 1983) indicates that phytoplankton and/or their microbial degraders may have interacted with copper in the lake water.

The Nonesuch deposits show organic matter had been buried to petroleum-generating temperatures, and Robbins (1985) made the suggestion that catalytic copper metallo-proteins may have entered into petroleum-generating reactions. Copper concentrations in heavy petroleum and bitumen in the region assay 2–4x higher than in surrounding reservoir rocks and sediments (Hosterman et al. 1990), indicating either higher bioaccumulation and/or greater retention of copper depending on local environmental conditions.

Beginning about 2 million years ago, continental glaciers, perhaps 1000 m thick, covered this northern landscape, scraped the land, redeposited boulders containing native copper (Reed,1991), and removed evidence of any interactions between Pleistocene organisms and copper.

Chalcocite ore, found within the lower beds of the Nonesuch shale, with an average concentration of 1.1% Cu has been mined since the late 1800s until 1995 at the White Pine Mine, Michigan (White and Wright 1954). Today brine pools in the White Pine Mine are coated with films of floating blue-green copper minerals and petroleum that drip through roof bolts 820 m underground. The green color of the films is due to the copper chloride minerals, atacamite and para-tacamite [$Cu_2Cl(OH)_3$], which surround Gram negative bacterial rods and filaments, leading Robbins et al. (1994) to suggest that modern bacteria participate in the mineral formation from mine water that has 7 ppm Cu. These minerals are enriched in C^{13} relative to the local petroleum, which is the likely carbon source. The bacterial role in forming the copper chloride minerals is presently under study.

Table 17.2: Geologic history, copper mineralization, and fossils in the Lake Superior Province

Age (billion years)	Geologic Time	Cu Mineralization	Fossils	References
3.9–4.6	Hadean	Cu in 66% meteorites		Rubin 1994
2.6–2.75	L. Archaean	Chalcopyrite in Atikokan area	Stromatolites with bacteria/cyanobacteria	Morey and Sims 1996; Kirkham 1979
		Native Cu, Soudan Iron Fm	Oldest organic C in Lake Superior range (bacteria?, cyanobacteria?)	Cloud, et al. 1965; Morey and Sims 1996
~1.9–2.1	M. Proterozoic	Carbonate-hosted stratiform Cu mineralization	Organic C and stromatolite fossils; algal filaments	Cloud, et al. 1965; Morey and Sims 1996
1.0–1.2		Volcanic and sedimentary hosted Cu deposits; native Cu along faults	Cyanobacterial sheaths, fungal hyphae-sized tissues, algae, and amorphous organic tissues; pellet-shaped microfossils	Alyanak & Vogel 1974; Cannon 1992; Cannon et al. 1989; Morey and Sims 1996; Nicholson, et al. 1992; Robbins 1983
0–0.6	Cambro-Ordovician	Cu oxide minerals stain cliff faces at Pictured Rocks National Lakeshore	Trilobite, brachiopods, mollusks, trace fossils	Runkel, et al. 1998; Wirth, et al. 1998
0.2	L. Ordovician to M. Devonian	None reported, but Noranda Cu deposit in Quebec of equivalent age	Bivalves, gastropods, conodonts, brachiopods, crinoids	Cannon and Nicholson 2001

Impact of the Last 10,000 Years

Archaic Indians arrived in the region about 10,000 years ago; they dug thousands of copper quarry pits, some several meters deep, on Isle Royale and on the Keweenawan Peninsula. Copper implements, shown to originate from the Lake Superior region by neutron activation analyses, have been found along the east coast of the United States as far south as Florida and even into Mexico (Julig et al. 1992).

Pre-European vegetation in the Lake Superior basin was characterized as a conifer-hardwood forest (Wright 1972). Where copper-bearing rocks weather at the surface, conifers can concentrate copper in tissues (to 700 ppm) and exudates (to 1500 ppm) above background soil (70 ppm) (Curtin et al. 1974). At Pictured Rocks National Lakeshore, Shacklette (1967) measured 1.5 weight % Cu in the ash of the copper-accumulating moss, Mielichhoferia, which was growing on a soil derived from a Cambrian sandstone (150 ppm Cu). Studies on phyto-uptake of copper show that "non-accumulator" terrestrial vegetation has mechanisms for excluding copper, unless the soil is acid (Chaney et al. 2000). Wild rice in non-contaminated lakes in the Lake Superior region naturally concentrates copper in edible seeds (5.3 ppm), in stems (1.4 ppm) and roots (4.8 ppm) (Bennett et al. 2000).

Toxicological studies undertaken in Lake Superior showed elevated copper in fish (Lucas et al. 1970), amphipods (Kraft 1979), mollusks (Kraft and Sypniewski 1981), and oligocheates (Phipps et al. 1995). The LC_{50} (lethal concentration for 50% of the test group) for fish is less than 1 µg Cu/L, as Cu above this concentration becomes overload and interferes with gill function and acid-base regulation (Tao et al. 2000, Wang et al. 1998). Fish and wild rice are staples of the Ojibwa subsistence populations, and there is the fear that copper accumulations might reach concentrations toxic to these humans (Joseph M. Rose, Sr., personal communication 1993).

Past and present industrial practices add to this concern: copper smelting emissions have been linked to morbidity as well as to respiratory cancers, although metals associated with copper, such as As, Cd, Ni, and Pb, are more usually implicated (Andrzejak et al. 1993, Hwang et al. 1997, Lubin et al. 2000). However, a case study in northern Ontario among Ojibwa-Cree Indians identified 6 children with high hepatic copper concentrations (Phillips et al. 1996).

Conclusions

Copper is readily bioavailable in the Lake Superior region. Though essential for life, elevated concentrations of copper can become toxic. As in the geosphere, biological copper is associated with other metals, such as As, Co, Fe, Mo, Ni, Pd, Pt, and Zn, each with toxicity thresholds and synergisms of their own. Additional concentrations of copper and other metals can be released to the environment from anthropogenic sources including mining, atmospheric deposition, copper plumbing (Eife et al. 1999), and cuprous chloride from road salt. (G.L. LaBerge, oral communication 2001). A major concern is that additional bioavailable concentrations of Cu may increase above toxicity thresholds of the biota living in the area, some of which are the major food staples of indigenous Americans.

Acknowledgments

We would like to thank Bill Cannon, Rufus Chaney, Howard Evans, Gunter Faure, Corey Goetsch, Cheryl Gullett-Young, Arthur Iberall, Gene LaBerge, Suzanne Nicholson, Doug Owens, Joseph Rose, Mike Sanders, Moto Sato, Klaus Schulz, Bob Seal, Terry Slonecker, Mark Stanton, Paul Strothers, Jochen Tilk, and Pam Winsky for help with field work, lab work, translation, or sharing ideas and references.

References

Andrzejak, R., Antonowic, J., Tomczyk, J., Lepetow, T., and Smolik, R., 1993, Lead and cadmium concentrations in blood of people living near a copper smelter in Legnica, Poland: Sci. Total Environ., Suppl. Pt. 1, p. 233–236.

Beck, M.T., and Ling, J., 1977, Transition-metal complexes in the prebiotic soup: Naturwissenschaften, v. 64, p. 91.

Bennett, J.P., Chiriboga, E., Coleman, J., and Walter, D.M., 2000, Heavy metals in wild rice from northern Wisconsin: Science of the Total Environment, v. 246, p. 261–269.

Cannon, W.F., 1992, The Midcontinent rift in the Lake Superior region with emphasis on its geodynamic evolution: Tectonophysics, v. 213-p. 41–48.

Cannon, W.F., and McGervey, T.A., 1991, Map showing mineral deposits of the Midcontinent rift, Lake Superior Region, United States and Canada: U.S. Geological Survey Map MF-2153.

Cannon, W.F., and Nicholson, S.W., 2001, Geologic map of the Keweenaw Peninsula and adjacent area, Michigan: U.S. Geological Survey Map I-2696.

Chandler, V.W., Bowman, P.L., Hinze, W.S. and O'Hara, N.W., 1982, Long-wavelength gravity and magnetic anomalies in the Lake Superior region, in Geology and Tectonics of the Lake Superior Region, eds. Wold, R.J., and Hinze, W.S., p 223–237.

Chaney, R.L., Li, Y.-M., Angle, J.S., Baker, A.J.M., Reeves, R.D., Brown, S.L., Homer, F.A., Malik, M., and Chin, M., 2000, Improving metal hyperaccumulator wild plants to develop commercial phytoextraction systems: Approaches and progress, in Terry, N., and Banuelos, G.S., eds., Phytoremediation of Contaminated Soil and Water, CRC Press, Boca Raton, p. 131–160.

Clark, J.L., 1974, Distribution and mode of occurrence of copper sulfides in the Kona Dolomite Marquette County, Michigan: Proceedings and Abstracts, Inst. Lake Superior Geology, Annual Meeting, v. 20, p. 9.

Cloud, P.E., Jr., Gruner, J.W., and Hagen, H., 1965, Carbonaceous rocks of the Soudan Iron Formation (early Precambrian): Science, v. 148, p. 1713–1716.

Cloud, P., and Morrison, K., 1980, New microbial fossils from 2 Gyr old rocks in northern Michigan: Geomicrobiology Journal, v. 2, p. 161–178.

Cowan, J.A., 1998, Copper proteins: http://bmbsgi11.leeds.ac.uk/promise/CUMAIN.html

Curtin, G.C., King, H.D., and Moser, E.L., 1974, Movement of elements into the atmosphere from coniferous trees in subalpine forests of Colorado and Idaho: Jour. Geochem. Explor., v. 3, p. 245–263.

Daniels, P.A., Jr., 1982, Upper Precambrian sedimentary rocks; Oronto Group, Michigan-Wisconsin, in Wold, R.J., and Hinze, W.J., eds., Geology and Tectonics of the Lake Superior Basin: Geol. Soc. America Mem. 156, p. 107–133.

Dexter-Dyer, B., Kretzschmar, M., and Krumbein, W.E., 1984, Possible microbial pathways in the formation of Precambrian ore deposits: Journal Geological Society, v. 141, p. 251–262.

Edhorn, A.-S., 1973 Futher investigations of fossils from the Animikie, Thunder Bay, Ontario: Proceedings Geological Assoc. Canada, v. 25, p. 37–66.

Eife, R., Weiss, M., Barros, V., Sigmund, B., Goriup, U., Komb, D., Wolf, W., Kittel, J., Schramel, P., and Reiter, K., 1999a, Chronic poisoning by copper in tap water: II. Copper intoxications with predominantly gastrointestinal symptoms: Eur. J. Med. Res., v. 4, p. 219–223.

Eisler, R., 2000, Copper, in Handbook of Chemical Risk Assessment: health hazards to humans, plants, and animals, Vol. 1 Metals, ed. R. Eisler, Lewis Publ., p. 93–200.

Fabiasiak, J.P., Pearce, L.L., Borisenko, G.G., Tyhurina, Y.Y., Tyurin, V.A., Razzack, J., Lazo, J.S., Pitt, B.R., and Kagan. V.D., 1999, Bifunctional anti/prooxidant potential of metallothionein: redox signaling of copper binding and release, Antioxid Redox Signal, v. 1, no. 3, p. 394–64.

Han, T.-M., and Runnegar, B., 1992, Megasopic eukaryotic algae from the 2.1-billion-year-old Negaunee Iron-Formation, Michigan: Science, v. 257, p. 232–235.

Hem, J.D., 1985, Study and interpretation of the chemical characteristics of natural water, 3rd ed.: U.S. Geological Survey Water-Supply Paper 2254, 263 p.

Honjo, S., Manganini, S.J., and Cole, J.J., 1982, Sedimentation of biogenic matter in the deep ocean: Deep-Sea Research, Part A, Oceanographic Research Papers, v. 29, p. 609–625.

Hosterman, J.W., Meyer, R.F., Palmer, C.A., Doughten, M.W., and Anders, D.E., 1990, Chemistry and mineralogy of natural bitumens and heavy oils and their reservoir rocks from the United States, Canada, Trinidad and Tobago, and Venezuela, USGS Circular 1047, pp. 19, 9 tables, 24 ref.

Hwang, Y.H., Bornschein, R.L., Grote, J., Menrath, W., and Roda, S., 1997, Environmental arsenic exposure of children around a former copper smelter site: Environ. Res., v. 72, p. 72–81.

Jacobsen, J.B.E., 1975 Copper deposits in time and space: Minerals Sci. Engineering, v. 7, p. 337–371.

Julig, P.J., Pavlish, L.A., and Hancock, R.G.V., 1992, Source determination of chert and copper artifacts by Neutron Activation Analysis: Some examples from the Great Lakes region: Geological Society of America, Annual Meeting, Cincinnati, Ohio.

Kirkham, R.V., 1979, Copper in iron formation: Geol. Surv. Canada, Paper no. 79–1B, p. 17–22.

Kraft, K.J., 1979, Pontoporeia distribution along the Keweenaw shore of Lake Superior affected by copper tailings: Journal of Great Lakes Research, v. 5, p. 28–35.

Kraft, K.J., and Sypniewski, R.H., 1981, Effect of sediment copper on the distribution of benthic macroinvertebrates in the Keweenaw waterway: Jour. Great Lakes Research, v. 7, p. 258–263.

LaBerge, G.L., 1994, Geology of the Lake Superior Region: Phoenix, Geoscience Press, 313 p.

Lubin, J.H., Pottern, L.M., Stone, B.J., and Fraumeni, J.F., 2000, Respiratory cancer in a cohort of copper smelter workers: Results from more than 50 years of follow-up: Am. J. Epidemiol., v. 15, p. 554–565.

Lucas, H.F., Edgington, D.N., and Colby, P.J., 1970, Concentrations of trace elements in Great Lakes fishes: Journal Fisheries Research Board of Canada, v. 27, p. 677–684.

Margulis, L., and Schwartz, K.V., 1998, Five Kingdoms: An illustrated guide to the phyla of life on earth, 3rd ed., W.H. Freeman and Co., 520 p.

Morey, G.B., and Sims, P.K., 1996, Metallogeny, in Sims, P.K., and Carter, L.M.H., eds., 1996, Archean and Proterozoic Geology of the Lake Superior Region, U.S., 1993: U.S. Geological Survey Professional Paper 1556, p. 89–95.

Mudrey, M.G., Jr., and Kalliokoski, J., 1993, Metallogeny, in Sims, P.K., et al., eds., The Lake Superior region and trans-Hudson orogen, Chapter 2, in Reed, J.C., Jr., et al., eds., Precambrian : Conterminous US: The Geology of North America, V. C-2: Boulder, Colo., Geological Society of America, p. 89–93.

Muller, T., Muller, W., and Feichtinger, H., 1998, Idiopathic copper toxicosis: Am J Clin Nutr, Suppl., v. 67, p. 1082S–1086S.

National Research Council, 2001 (in press), Dietary Reference Intakes for Vitamin A, Vitamin K, Arsenic, Boron, Chromium, Copper, Iodine, Iron, Manganese, Molybdenum, Nickel, Silicon, Vanadium, and Zinc, National Academy Press, 650 p.

Nicholson, S.W., Cannon, W.F., and Schulz, K.J., 1992, Metallogeny of the Midcontinent rift system of North America: Precambrian Research, v. 58, p. 355–386.

Phillips, M.J., Ackerley, C.A., Superina, R.A., Roberts, E.A., Filler, R.M., and Levy, G.A., 1996b, Erratum: Excess zinc associated with severe progressive cholestasis in Cree and Ojibway children: Lancet, v. 347, p. 1776.

Phipps, G.L., Mattson, V.R., and Ankley, G.T., 1995, Relative sensitivity of three freshwater benthic macroinvertebrates to ten contaminants: Archives of Environmental Contamination and Toxicology, v. 28, p. 281–286.

Pourbaix, M., 1966, Atlas of electrochemical equilibria in aqueous solutions: New York, Pergamon, 644 p.

Reed, R.C., 1967, Copper mineralization in Animikie sediments of the eastern Marquette Range, Marquette County, Michigan: Compass of Sigma Gamma Epsilon, v. 45, p. 47–55.

Reed, R.C., 1991, Economic geology and history of metallic minerals in the Northern Peninsula of Michigan, in Catacosinos, P.A., and Daniels, P.A., Jr., eds, Early Sedimentary Evolution of the Michigan Basin: Geol. Soc. America Special Paper 256, p. 13–51.

Robbins, E.I., 1983, Accumulation of fossil fuels and metallic minerals in active and ancient rifts: Tectonophysics, v. 94, p. 633–658.

Robbins, E.I., 1985, Petroleum as an ore-bearing fluid: A hypothesis [abs.]: American Association of Petroleum Geologists, NE Section meeting, Williamsburg, VA, American Association of Petroleum Geologists Bulletin, v. 69, no. 9, p. 1446.

Robbins, E.I., Porter, K.G., and Haberyan, K.A., 1985, Pellet microfossils: Possible evidence for metazoan life in Early Proterozoic time: Proc. Nat. Acad. Sci., v. 82, p. 5809–5813.

Robbins, E.I., Stanton, M.R., Tilk, J.E., Congdon, R.E., Evans, H.T., Jr., Gullett, C.D., Sanders, M.B., Sato, M., Schaef, H.T., and Seal, R.R., II, 1994, Association of microbes with authigenic copper-chloride mineral films (atacamite and paratacamite) and petroleum residue at depth in the White Pine copper mine (abs.): Waterloo '94, Program with Abstracts, Geological Assoc. Canada and Mineralogical Assoc. Canada, Annual Meeting, p. A94.

Robbins, E.I., and Traverse, Alfred, 1980, Degraded palynomorphs from the Dan River (North Carolina)-Danville (Virginia) basin, in Price, V., Jr., Thayer, P.A., and Ranson, W.A., eds., Geological Investigations of Piedmont and Triassic Rocks, Central North Carolina and Virginia, Carolina Geological Society Field Trip Guidebook, p. BX 1–11.

Rubin, A.E., 1994, Metallic copper in ordinary chondrites: Meteoritics, v. 29, p. 93–98.

Runkel, A.C., McKay, R., Palmer, A.R., 1998, Origin of a classic cratonic sheet sandstone; stratigraphy across the Sauk II-Sauk III boundary in the Upper Mississippi Valley: Geological Society of America Bull., v. 110, p. 188–210.

Shacklette, H.T., 1967, Copper mosses as indicators of metal concentrations: U.S. Geological Survey Bulletin 1198-G, p. G1–G18.

Shields, W.R., Goldich, S.S., Garner, E.L., and Murphy, T.J., 1965, Natural variations in the abundance ration and the atomic weight of copper: Jour. Geophys. Research, v. 70, p. 479–491.

Sillitoe, R.H., Folk, R.L., and Saric, N., 1996, Bacteria as mediators of copper sulfide enrichment during weathering: Science, v. 272, no. 5265, p. 1153–1155.

Sims, P.K., and Carter, L.M.H., eds., 1996, Archean and Protero-zoic Geology of the Lake Superior Region, U.S., 1993: U.S. Geological Survey Professional Paper 1556, 115 p.

Sternlieb, I., 2000, Wilson's disease: Clin. Liver Dis., v. 4, p. 229–239.

Strauss, H., 1986, Carbon and sulfur isotopes in Precambrian sediments from the Canadian Shield: Geochim. Cosmochim. Acta, v. 50, p. 2653–2662.

Strother, P.K., 1986, Palynomorphs from the copper-bearing Nonesuch Formation (abs.): Palynology, v. 19, p.35.

Tao, S., Long, A., Liu, C., and Dawson, R., 2000, The influence of mucus on copper speciation in the gill microenviron-ment of carp (Cyprinus carpio): Ecotoxicol. Env. Saf., v. 47, p. 59–64.

Walker, E.C., Cuttitta, F., and Senftle, F.E., 1958, Some natural variations in the relative abundance of copper isotopes: Geochimica et Cosmochimica Acta, v. 15, p. 183–194.

Wang, T., Knudsen, P.K., Brauner, C.J., Busek, M., Vijayan, M.M., and Jensen, F.B., 1998, Copper exposure impairs intra- and extracellular acid-base regulation during hypercapnia in the fresh water rainbow trout (Oncorhynchus mykiss): J. Comp. Physiol [B], v. 168, p. 591–599.

Weiler, R.R., 1978, Chemistry of Lake Superior: Jour. Great Lakes Res., v. 4, p. 70–385.

Weissman, Z., Berdicevsky, I., Cavari, B.Z., Kornitzer, D., 2000, The high copper tolerance of Candida albicans is mediated by a P-type ATPase, Proc Natl Acad Sci U.S., v. 97, no. 7, p. 3520–5.

White, W.S., and Wright, J.C., 1954, The White Pine copper deposit, Ontonagon County, Michigan: Economic Geology, v. 49, p. 675–716.

Wilks, M.E., and Nisbet, E.G., 1985, Archaean stromatolites from the Steep Rock Group, northwestern Ontario, Canada: Canadian Journal of Earth Sciences, v. 22, p. 792–799.

Wirth, K.R., Cordua, W.S., Kean, W.F., Middleton, M., and Naiman, Z.J., 1998, Field guide to the geology of the southwestern portion of the Midcontinent rift system, eastern Minnesota and western Wisconsin: Field Trip 2, 44th Annual Meeting, Minneapolis, Minnesota, Institute on Lake Superior Geology, v. 44 Part 2, p. 57–59.

Wright, H.E., Jr., 1972, Physiography of Minnesota, in Sims, P.K., and Morey, G.B., eds., 1972, Geology of Minnesota, A Centennial Volume: St. Paul, Minnesota Geol. Survey, p. 561–578.

Zhu, X.K., O'Nions, R.K., Guo, Y., Belshaw, N.S., and Rickard, D., 2000, Determination of natural Cu-isotope variation by plasma-source mass spectrometry; implications for use as geochemical tracers: Chem. Geol., v. 163, p. 139–149.

18

Health Problems Related to Environmental Fibrous Minerals

Gunnar Hillerdal, M.D.
Departments of Lung Diseases
Karolinska Hospital, Stockholm, and Akademic Hospital, Uppsala, Swedens

Introduction

Very early in the history of mining and with the industrial applications of asbestos, it was known that a variety of mineral fibers were hazardous to health. By the 1940s, the potential risk of lung cancer, in addition to the fibrosis disorder asbestosis, one of the pneumoconioses, was described. Within twenty years, another malignant disease, mesothelioma, cancer of the tissues that surround the lung, was ascribed to asbestos exposure. It is now common knowledge that inhalation of certain mineral fibers can cause disease (Skinner et al. 1988). Because the fibers are inhaled, the lung and surrounding tissues are the primary targets, but there may be subsequent reactions in many other parts of the body. The information on disease related to fibrous materials emanates from studies of occupational environments where the dose or exposure is likely to be high and continued over long periods of time. However, it has been increasingly realized that domestic or general environmental exposure is also possible and can pose grave dangers.

For a mineral fiber to be inhalable it should be less than one micron in diameter, but the length can be 10 microns or greater because the particle can align with the air stream in the bronchi and penetrate far into the lung. The ratio between length and diameter of the fiber is critical. The most dangerous fibers are very thin (one tenth of a micron in diameter or less) with a high length-diameter ratio.

Another important factor is biodurability. Typically, the dangerous fibers are not broken down at all or only very slowly, with half-lives in the body of many years. They may remain in situ throughout life and can be found at autopsy.

There are many varieties of fibers in the environment today, both naturally occurring and man-made. Only a few, however, fulfill the above criteria and occur in amounts where human exposure is possible. The problem fibers are collectively known as asbestos and the fibrous zeolite, erionite. There are many other fibers (Skinner et al. 1988), but their contributions to human disease are not recognized.

Mineral Fibers of Medical Interest

Asbestos is not a mineralogical but a commercial term. There are six individual minerals classified as asbestos when they occur in fibrous form (Skinner et al. 1988). The chemically distinct minerals that form straight fibers are from the amphibole mineral group. The mineral names are crocidolite ("blue asbestos"), amosite ("brown asbestos"), tremolite, actinolite, and anthophyllite. However, the most common and widely used type is chrysotile, or white asbestos, which has

Table 18.1: Mineral fibers of medical interest

Name	Relative Size	Biodurability	Use	Diseases
ASBESTOS: AMPHIBOLES (straight fibers)				
Crocidolite	Very thin and long	Very high	Now little	Very high risk mesothelioma Lung Cancer, Plaques
Tremolite plaques	Thin, long	High	None Contaminant	High risk mesothelioma, lung cancer,
Amosite	As tremolite	High	Limited	As for tremolite
Anthophyllite	Thicker fibers	High	Now none	Very low risk mesothelioma Median risk lung cancer Very high incidence plaques
CHRYSOTILE (curly fibers)	Thin and short Can be very long	Fairly low	> 90% of commercial asbestos	Not certain that it can cause disease except lung cancer (see text)
ZEOLITES: Erionite	Extremely thin Long	High	None	Extremely high risk for mesothelioma; lung cancer unknown; plaques occur

curly fibers, and is a member of the serpentine mineral group. All these fibers may occur with different diameters and lengths, and also in their ability to resist breakdown in biological tissues (Table 18.1). Chrysotile has the fastest clearance from the body, and diseases seem to be mainly associated with exposure to the amphiboles (Churg 1982).

Whether chrysotile can cause mesothelioma has been hotly debated in the last few years. Because this asbestos variety has relatively short survival in the body, only rarely can one find high levels at autopsy even in chrysotile miners. The hypothesis is that one of the other amphibole asbestos varieties is a more likely culprit. Mesothelioma, an uncommon cancer, even in miners, has been associated with the amphibole group fibrous mineral crocidolite. The indictment that tremolite, a very common mineral in non-fibrous or blocky morphology, may contaminate commercial chrysotile ore, and be responsible for any disease, has not been proved. Fibrous, or asbestiform tremolite, occurs very rarely.

Crocidolite is the most dangerous of the asbestos fibers and is no longer mined. This fiber has a high resistance to acids, which made it a very useful industrial substance. Amosite is also rarely used nowadays. Tremolite has been mined only to a small extent but is a common contaminant in talc mines as well as chrysotile and also many other ores, such as nickel and iron. Tremolite is one of the most common asbestos minerals, occurring in many rocks all over the world. Anthophyllite has been mined, for example, in Finland and Japan, but today has no industrial use.

Erionite, finally, is not a very common mineral but is formed under certain conditions in volcanic areas of the world and is found with other zeolite minerals often in thick layers.

Diseases and Radiological Findings Caused by Asbestos and Erionite

These can be divided into malignant and benign. The first described was asbestosis, or pulmonary fibrosis, which can occur with all types of asbestos. The lung becomes fibrotic and stiff and gas exchange dramatically decreases. It is a dose-related disease, and a fairly high exposure is necessary to cause the clinical manifestations: the patient gets more and more short of breath. Once the process has started, it continues to worsen. However, asbestosis is rare with environmental exposure but can occur after many years of slight exposure.

The other benign lesions associated with asbestos are pleural plaques. In the 1950s, it became clear that plaques were the most common manifestation of asbestos inhalation. Macroscopically, they are shining white elevations with sharp borders on the inside of the chest wall and consist of fibro-hyaline connective tissue containing very few cells. There is no inflammation within the plaques, but small cellular aggregates can be seen at their periphery, indicating low-grade inflammation there.

Over 80% of strictly defined pleural plaques are due to occupational exposure to asbestos (Hillerdal 1978, Hillerdal 1994, Karjalainen 1994). In the general population there are no "endemic" plaques but the occasional appearance without clinical symptoms may occur in individuals who had only low-level or sporadic exposure.

The relation between dose and response for pleural plaques is much weaker than that for asbestosis. The mean number of asbestos fibers or bodies in persons with plaques is as a rule higher than in the normal population (Karjalainen et al. 1994, Kishimoto et al. 1989, Warnock et al. 1982), but there is a fairly large variation and a number of persons with plaques may have values little or no different from the general population. In other words, plaques are associated with a wide range of asbestos burdens that overlaps with that found in any control population.

Plaques are more related to time after exposure than to the dose (Hillerdal 1978). Very few plaques will be seen earlier than 15 years after the first exposure to asbestos, and most will appear only after 30 years. In areas where the population is exposed from birth, the first pleural changes appear after age 30 and the incidence then increases with age. Thus, occurrence of plaques is dependent on cumulative exposure and time since first exposure (Järvholm 1992). Many plaques are not

seen until long after exposure has ceased. Once seen, they will slowly grow larger over the years, and with time many will calcify (Hillerdal 1978).

The malignant diseases due to asbestos exposure are mainly lung cancer and malignant mesothelioma. Other tumors in the body have been claimed to be more abundant in asbestos workers, but the scientific proof varies and they can be disregarded in this review.

Lung cancer is mainly a disease caused by smoking. However, exposure to asbestos increases the risk in a dose-related manner. Most data point to a multiplicative effect. For example, it can be estimated that smoking 20 cigarettes a day will increase the risk of lung cancer at least 10 times. If this smoker has an "average" occupational exposure to asbestos – say, he has been a construction worker – the risk will double to 20 times. Thus, asbestos exposure will more likely result in smokers developing lung cancer. In a non-smoker exposed to asbestos, the risk of developing a lung cancer is very small. In other words, most "asbestos lung cancers" are also "smokers' cancers." Lung cancer is thus a difficult disease to use as a marker of low dose exposure to mineral fibers, since smoking is overwhelmingly the primary cause.

Malignant mesothelioma, a very rare cancer, has no correlation to smoking. Apart from a small basal incidence due to unknown causes (and some other, very rare causes such as radiation or old pleural scars), all cases can be considered to be due to exposure to asbestos (or erionite). The tumors grow slowly in the pleura, restricting the action of the lung and causing the death of the patient often within a year of diagnosis. No curative treatment exists, though some cytostatics seem to have effect on the tumor. The disease is dose-related (Hillerdal 1999), but even a slight exposure can be enough. Like pleural plaques, the appearance is usually more than 30 years after exposure. Rarely, the disease starts in the peritoneum.

Thus, two findings — pleural plaques and malignant mesothelioma — are useful as "sentinel" diseases, indicating that asbestos (or erionite) exposure has occurred.

Exposure to Mineral Fibers

Although the relationship of exposure leading to disease is usually occupational, there are the possibilities in domestic situations such as general pollution of the environment from industrial uses, buildings containing asbestos, and finally, and of most interest for this review, the local geologic occurrences of mineral fibers.

Domestic Exposure

The family of an asbestos worker could be exposed to considerable amounts of asbestos brought home in his working clothes, which all too often it was the duty of the wife and/or daughter to clean. Pleural plaques and pulmonary fibrosis have been described in such cases, as well as mesotheliomas (Epler et al.1980, Kilburn et al. 1985).

Environmental Pollution

Asbestos factories some decades ago used to be great polluters, and pleural plaques and mesotheliomas have been reported in the general populations in the surroundings (Hillerdal 1999).

Asbestos has been introduced into a variety of consumer goods, in most cases without the users knowing of the content. Examples are wall paints, papier maché, spackling and jointing compounds, hair dryers, and cigarette filters. Once asbestos is introduced into a home, it will spread to all rooms and is almost impossible to remove even with a vacuum-cleaner. It will easily be disturbed into the air from the slightest movement, and sedimentation is very slow (Moorcroft & Duggan 1984, Rohl et al. 1975).

Buildings Containing Asbestos

A mixture containing asbestos was popular after the Second World War for spraying on ceilings and walls for insulation and decoration. There is now a public danger because of natural tear and wear, vandalism, or "artistic" carving in schools. Asbestos was also extensively used in the construction of buildings, for roof tiles, insulation, and as a cheap filler. The risks of any health effects from non-friable asbestos in public buildings are nonexistent or extremely low and thus the cost of removal is not warranted (Whysner et al. 1994).

Local Deposits of Fibrous Minerals

In many areas of the world, asbestos fibers occur in the soil, remnants of rock weathering processes. Inasmuch as chrysotile is more soluble, the amphiboles are the fibers usually found. Farmers working with the soil may be exposed, and in some countries the locally occurring asbestos has been used for whitewashing of houses, construction of fireplaces, or sauna stones (Baris et al. 1988, Hillerdal 1997, Kiviluoto 1960, Meurman 1968, Puffer et al. 1980, Raunio 1966, Yazicioglu et al. 1980). As a result, there are areas of the world where pleural plaques are endemic. Endemic pleural plaques were first described from Finland and since then many other sites have been reported (Table 18.2).

Calcified plaques are detected radiographically in the older age groups and at autopsy (up to 100% in persons above age 50), though usually the incidence is more modest. Plaques are more common among male farmers whereas, when mineral fibers are used for whitewashing of houses, the women also have a high incidence of plaques.

One of the best described occurrences of fibrous diseases is Turkey, where the whitewash with tremolite was common in many villages. In addition, exposure to a fibrous zeolite called erionite also has been reported. This mineral was formed during volcanic activity and occurs locally in horizontal layers close by the best known exposed village of Karain. The erionite can be found in roads, fields, and building stones. Apart from the pleural changes, these villages also have an extremely high incidence of malignant mesothelioma. In fact, this dreadful disease is the main cause of death there (Baris 1981).

Endemic plaques are of interest also in other countries, since many persons born in "endemic areas" have moved to other places, taking with them the plaques and also the risk of mesothelioma and lung cancer.

What Should be Done about Asbestos and Erionite Occurring Locally?

As mentioned, Turkey is the best investigated country for these local findings, but as can be seen from Table 18.2, many other countries are also affected. Most probably, the problem is much more widespread and

Table 18.2: Local deposits of mineral fibers (asbestos or erionite), incidence of plaques, and of malignant mesothelioma (For references, see Hillerdal 1991)

Type of fiber	Country or area	Population or activity	Incidence of plaques in the endemic area	Incidence of mesothelioma
Tremolite				
	Austria	Vineyard & field workers	Increased	Not increased
	Bulgaria	Tobacco growers	Increased	Not increased
	Corsica	General population	Increased	Increased
	Cyprus	General population	Increased	Increased
	Greece	White-washing houses	High	Increased
	New Caledonia	White-washing houses	High	High
	Turkey	White-washing houses Farmers	High	High
Amosite				
	Africa	Population around mine	High	High
Crocidolite				
	South Africa	Population around mine	High	High
	P. R. China	General population	High	High
Anthophyllite				
	Finland	Population around mine	High	Not increased
	Japan	Population around mine	High	Not increased
Unknown				
	Czechoslovakia	Farmers	High	Unknown
	Former USSR	General population	Increased	Unknown
Erionite				
	Turkey	Villagers	High	Extremely high

other countries may have similar problems, but not detected or identified as yet. Corsica, Cyprus, Greece, and New Caledonia have only recently discovered these diseases in their populations (Hillerdal 1991) and in Turkey new villages are reported almost every year (Metintas 1999).

Once the problem has been identified, the most important next step is to inform those affected of the risk and which habits should be avoided. For instance,

whitewashing of houses should be abandoned or other substances should be used. Good results will be achieved by prevention method alone, as shown in a recent paper from Turkey and Greece where there are indications that the incidence of mesothelioma is going down (Constantopoulos et al. 1987).

The responsibility of the medical and geological experts working together to identify the disease and the problem mineral is to reduce disease. It can be difficult, as in the erionite case in Turkey, where the problem is not any particular use of the substance but rather that it occurs in the ground. The only solution is then to move the village. Sometimes things can go wrong, as in the example of Karain Village. The press, including of course the Turkish, obtained the story and headlines such as "Village of Death" were seen on articles worldwide. This caused not only villagers to have difficulties marrying outside the village but even such problems as people refusing to buy vegetables or other products from the area. Naturally, such problems will sour the relations with researchers. One must be careful and try to avoid such unintended results of any intervention. Cooperation between the inhabitants and the medical profession and officials is very important and it must never be forgotten that the goal is to help the population.

References

Baris YI, Saracci R, Simonato L et al. Malignant mesothelioma and radiological chest abnormalities in two villages in central Turkey. An epidemiological and environmental investigation. Lancet 1981; I, p.984–987.

Baris YI, Bilir N, Artvinli M et al. An epidemiological study in an Anatolian village environmentally exposed to tremolite asbestos. Brit J Industr Med 1988; v.45, p.838–840.

Churg A. Fiber counting and analysis in the diagnosis of asbestos-related disease. Hum Pathol 1982; v.13, p.381–392.

Constantopoulos SH, Saratzis N, Kontogiannis D et al. Tremolite whitewashing and pleural calcifications. Chest 1987; v.92, p.709–712.

Epler GR, Gerald MXF, Gaensler A, Carington CB. Asbestos-related disease from house-hold exposure. Respiration 1980; v.39, p.229–240.

Hillerdal G. Pleural plaques in a health survey material. Frequency, development, and exposure to asbestos. Scand J Respir Dis 1978; v.59, p.257–263.

Hillerdal G. Pleural plaques and risk for bronchial carcinoma and mesothelioma. A prospective study. Chest 1994; v.105, p.144–50.

Hillerdal G. Pleural plaques: Incidence and epidemiology, exposed workers and the general population. A review. Inddor Built Environ 1997; v.6, p.86–95.

Hillerdal G. Mesothelioma: casdes associated with non-occupational and low-dose exposures- Occup Environ Med 1999; v.56, p.505–13.

Järvholm B. Pleural plaques and exposure to asbestos: a mathematical model. Int J Epidemiol 1992; v.21, p.1180–1184.

Karjalainen A, Karhunen PJ et al. Pleural plaques and exposure to mineral fibers in a male urban necropsy population. Occup Environ Med 1994; v.51, p.456–460.

Kilburn KH, Lilis R, Anderson H et al. Asbestos disease in family contavts of ship yard workers. Am J Public Health 1985; v.75, p. 615–617.

Kishimoto T, Ono T, Okada K, Ito H. Relationship between number of asbestos bodies in autopsy lung and pleural plaques on chest X-ray film. Chest 1989; v.95, p.549–552.

Kiviluoto R. Pleural calcification as a roentgenologic sign of non-occupational endemic anthophyllite asbestosis. Acta Radiol 1960; Suppl 194.

Metintas M, Özdemir N, Hillerdal G et al. Environmental asbestos exposure and malignant pleural mesothelioma. Respir Med 1999; v.93, p.349–55.

Meurman LO. Pleural fibrocalcific plaques and asbestos exposure. Environ Res 1968; v.2, p.30–46.

Moorcroft JS, Duggan MJ. Rate of decline of asbestos fibre concentration in room air. Ann Occup Hyg 1984; v.28, p.453–457.

Puffer JH, Germine M, Hurtubise DO et al. Asbestos distribution in the central serpentine district of Maryland - Pennsylvania. Environ Res 1980; v.23, p.233–246.

Raunio V. Occurrence of unusual pleural calcification in Finland. Studies on atmospheric pollution caused by asbestos. Ann Med Int Fenniae 1966; v.55 Suppl. 47.

Rohl AN, Langer AM, Selikoff IJ, Nicholson WJ. Exposure to asbestos in the use of consumer spackling, patching, and taping compounds. Science 1975; v.189, p.551–553.

Skinner, H.C.W., Ross, Malcolm and Frondel, Clifford, 1988, Asbestos and other fibrous materials: mineralogy, crystal chemistry and health effects. Oxford University Press., New York.

Warnock ML, Prescott BT, Kuwahara TJ. Numbers and types of asbestos fibers in subjects with pleural plaques. Am J Pathol 1982; v.109, p.37–46.

Whysner J. Covello VT, Kuschner M et al. Asbestos in the air of public buildings: a public health risk? Prev Med 1994; v.23, p.119–125.

Yazicioglu S, Ilcayto R, Balci B et al. Pleural calcification, pleural mesotheliomas and bronchial cancers caused by tremolite dust. Thorax 1980; v.35, p.564–569.

19

Anthropogenic Distribution of Lead

H.W. Mielke, C. Gonzales, E. Powell, Sabrina Coty, and Aila Shah
College of Pharmacy, Xavier University of Louisiana, New Orleans, Louisiana

Introduction

In 1980, Clair C. Patterson made an astonishing statement: "Sometime in the near future it probably will be shown that the older urban areas of the United States have been rendered *more or less uninhabitable* [emphasis added] by the millions of tons of poisonous industrial lead residues that have accumulated in cities during the past century" (NAS 1980). This chapter describes the anthropogenic redistribution of Pb during the twentieth century from mineral-rich geologic formations to urban locations. It explores the meaning of "more or less uninhabitable" in the context of medical knowledge and presents empirical evidence about the neurotoxic impact of Pb to urban inhabitants, especially children.

Lead Redistribution:
From Mines to Urban Environments

Of the many lead- (Pb) containing products, Pb-based paint and Pb additives to gasoline contributed the largest quantities of Pb within urban environments. In the U.S., the gross tonnage of white-Pb used as pigment in paint from the late nineteenth century to 1978 was about equal to the amount used in tetra ethyl (and methyl) lead (TEL/TML) additives to gasoline be-

tween 1929 and 1986 (Mielke and Reagan 1998). The maximum uses of each of these Pb products occurred in two peaks. The first and lower peak occurred in the 1920s during the use of Pb-based paint when the U.S. economy was about to switch from agrarian to industrial. At that time, rail transportation was the main conveyance for moving goods and providing services, and Pb-based paints were applied to homes in both large and small communities throughout the nation. Pb-based paints remain as thin layers that, when intact, are not bioavailable. However, deterioration or careless removal of Pb-based paint contributes dust that contaminates the local environment with bioavailable Pb (Mielke et al. 2001).

The second and higher peak of Pb use occurred with leaded gasoline around 1970 at a time when the U.S. economy was industrial, urban, and relied on automobiles for transportation. In contrast to Pb-based paint coatings, combustion of Pb-gasoline emits particles of dust that are extremely bioavailable. Consequently, the processes that distributed Pb from gasoline are tied to two features of modern cities. First, most U.S. cities have a ground transportation system dominated by private automobiles and a highway network that concentrates traffic into the city. Second, the automobile users distributed the Pb contained in the fuel they consumed while driving. The emission of fine aerosol particles of Pb became concentrated in the most densely populated areas of U.S. cities (Mielke and Reagan 1998).

The 5.9 million metric tons of Pb in gasoline left 4–5 million metric tons of Pb in the environment (see Mielke and Reagan 1998). The difference between total and emitted Pb was physically trapped in the motor oil and the exhaust system of the cars, which may be deposited to city landfills, local soils, and waterways. Traffic flow maps provide the data for evaluating the impact of leaded gasoline in the large city of New Orleans and a neighboring Louisiana small town, Thibodaux. During 1968–1972, when Pb in gasoline was at its peak, the average daily vehicle traffic (ADVT) of 95,000 generated 5.15 metric tons at busy intersections in New Orleans (Mielke et al. 1997). This amount is compared to around 0.45 metric tons emitted at the busiest intersections in Thibodaux with an ADVT of 10,000. Soil Pb concentrations in communities around busy intersections of New Orleans are 300–1200 mg/kg, compared to 60 mg/kg near the busiest intersections of Thibodaux (Mielke et al. 1997).

Aerosol deposition of Pb and other metals are integrated into soil and become a relatively stable reservoir of contamination. Studies in Maryland, Minnesota, and Louisiana demonstrate a consistent pattern of soil Pb in urban environments that are related to city size and community location (Mielke 1993, Mielke et al. 1983, 1984). Soil Pb concentrations diminish with distance from the city center. For example, in Baltimore, the largest contamination (500–1000 mgPb/kg) is clustered in residential neighborhoods around the city center, less contamination occurs in suburban areas (50–200 mgPb/kg), and smallest concentrations occur in rural areas (10–25 mg Pb/kg). The probability that the inner-city clustering of Pb could be due to chance is one in 10^{23} (Mielke et al. 1983). Soils in old communities of large cities have Pb concentrations one or two orders of magnitude higher than similarly aged communities in small cities (Mielke et al. 1984–5, 1989, Mielke 1993). Potentially hazardous amounts of Pb were deposited in population centers of all major U.S. cities.

Medical Consequences of Pb

Lead poisoning due to the ingestion or inhalation of inorganic lead compounds is a widely recognized public health problem (CDC 1991). Clinically, blood-lead concentration is used to measure Pb exposure. In the general population, the highest risk of Pb poisoning occurs among children because of the developmental stage of their neurological system. Blood-lead concentrations as small as 10 to 15 µg/dL have been associated with neurological impairment of children (CDC 1991), and larger blood-lead concentrations are inversely correlated with performance on standardized intelligence tests (Mushak et al. 1989). Adverse health effects such as impaired hearing acuity and interference with vitamin D metabolism have also been observed at blood-lead levels of 10 to 15 µg/dL (CDC 1991). The Pb exposure of children appears to persist in the form of a variety of health problems throughout life (Dietert et al. 2000, Mushak et al. 1989). For chronic exposures of 10–15 mg/dL, Pb is a significant cofactor of decreased learning ability and delinquent behavior in later life (Needleman et al. 1996).

On January 1, 1986, the U.S. banned Pb additives in gasoline for highway use and this resulted in a remarkable reduction in the prevalence of childhood Pb exposure. In the mid-1970s, 9 out of 11 U.S. children (around 15 million children) were exposed to blood Pb concentrations of 10 mg/dL or more. In the late 1980s, after the ban, 1 out of 11 U.S. children (slightly less than 1 million children) were exposed to blood Pb concentrations of 10 mg/dL or more (Pirkle et al. 1994). Pb poisoning remains endemic mainly among African-American and other minority children who live in the inner-city areas of major U.S. cities (Brody et al. 1994).

One common route of exposure is through hand contact with Pb-contaminated dust and soil and subsequent ingestion by hand-to-mouth similar to pica behavior (**Abrahams**). In New Orleans, the Pb concentration on children's hands was studied at daycare centers in various parts of the city. In the inner-city an average of approximately 30 mg of Pb was measured on children's hands after outdoor play, or six times more Pb than they obtain while playing indoors (Viverette et al. 1996). The ease of chronic poisoning takes place because children's hands offer five times more Pb than the 6 mg per day total tolerable daily intake (TTDI) of Pb for children less than 6 years (Bourgoin et al. 1993). In mid-city New Orleans, smaller amounts (5 mg Pb per hand) were picked up both indoors as well as outside. The amounts of Pb on children's hands are directly related to the amounts of Pb measured in the outdoor soil (Viverette et al. 1996).

In New Orleans the association (Fisher's exact test $P = 3.18 \times 10^{-24}$) between soil Pb and blood Pb is significant and (Mielke et al. 1997). Least sum of absolute deviations regression were modeled and the resulting equation for the model is BL = 1.845 + 0.7215 (SL) $^{0.4}$ where BL is blood Pb level and SL is soil Pb concentration (Mielke et al. 1999). The correlation between observed and predicted blood Pb values is $r^2 = 0.552$, and the chance that predicted blood Pb is a "random fit" of the observed blood Pb is very remote ($P = 2.83 \times 10^{-23}$). A soil Pb map is a better gauge for predicting childhood Pb exposure than a map of age of housing, because old paint is only one source of the Pb exposure problem. Few cities have been mapped for Pb and other metals to the extent that New Orleans has (see Mielke 1994, Mielke et al. 2000), and therefore health care providers have limited information about the geochemical conditions that confront them.

Consequences of Urban Pb for Children

About 30% of the children 6 years old and younger living in New Orleans exhibit blood Pb levels 10 mg/dL or higher (Mielke et al. 1997). Exposure is not evenly distributed throughout the city. Exposure is endemic to old neighborhoods (in association with Pb painted houses) that also have a long history of abundant traffic flows. As discussed above, a major clinical consequence of Pb exposure is learning problems. Because the New Orleans school system is organized into attendance districts whereby schools draw students from the surrounding neighborhood, it is possible to examine the relationship between Pb in a community and children's learning performance. Standardized test results given to all 4th grade students were used to select two attendance districts for the study. Both elementary schools were predominantly attended by African-American students (> 98%). Table 19.1 summarizes the learning characteristics of the schools. Note that the 23.6 performance score for school A is

Table 19.1: Comparison of school characteristics and neighborhood characteristics for the case study of two New Orleans elementary school districts, A and B

School characteristics	A	B
Students-teacher ratio	14.8	25.8
Attendance rate %	94.1	95.9
Class size; %>20 pupils	100	83
Performance score [a]	23.6	74.8
% regular education	89	99
% special education	11	1
Number of students	322	685

(a) School Performance Score is calculated by the New Orleans School system from index results of 3 measures: LEAP scores, the Iowa Tests, and attendance rates for regular students only.

Table 19.2: Comparison of the neighborhood soil characteristics for the case study of two New Orleans elementary school districts, A and B

Neighborhood Characteristics		A	B
Soil Pb (mg/kg)			
Percentiles	Min	5	16
	10%	37	37
	25%	200	88
	Median	405	152
	75%	1040	363
	90%	1891	437
	Max	21679	938
Mean		3608	290
N=		26	26

Fisher-Pitman Permutation test *P*-value 0.015

significantly below 74.8 for school B where the record is good for learning performance.

The U.S. Environmental Protection Agency (EPA) states that a hazard exists when Pb is greater than 400 mg/kg (ppm) in bare soil for children's play areas, or 1200 mg/kg average for bare soil in the rest of the yard

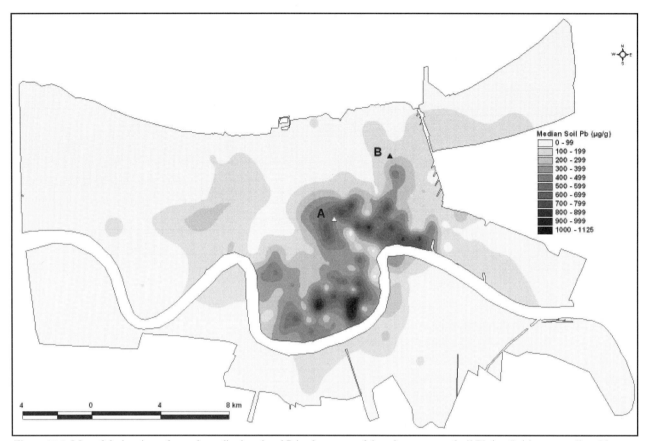

Figure 19.1: Map of the locations of attendance districts A and B in the context of the urban pattern of soil Pb (mg/kg) in metropolitan New Orleans (For further information about the survey, see Mielke et al. 1999).

(USEPA 2001). The soil samples for this project were collected from areas with common access by all children living in the respective attendance district. Table 19.2 provides a description of the soil Pb results from attendance districts A and B. Note that at 3608+mg Pb/kg the average soil Pb of attendance district A exceeds the EPA 1200 mg/kg standard while the mean for Pb in B (290 mg/kg) does not. In addition, in attendance district A at least half of the soils collected were above 400 mg Pb/kg while about 10% of the soils in B were above that standard. Figure 19.1 shows the location of schools A and B and the predicted pattern of median soil Pb in New Orleans. Note that the results obtained for the case study of the two attendance district schools was within the predicted range (Mielke et al. 1999). Given what is known about the association between soil Pb and children's exposure, as well as Pb exposure and learning, chronic Pb poisoning of preschool children is probably endemic in attendance district A. From the perspective of learning potential, we suggest that Pb may be one factor in the low performance scores. Confounding factors such as family

income, marital status, and educational attainment of parents may also contribute to the differences between A and B (Mielke et al. 1999). The results for New Orleans are similar to those of a study in Ohio where Pb was also implicated as a major risk factor for successful learning of youth in urban compared to suburban schools (Boyd et al. 1999).

Geologic and earth science agencies have personnel and facilities that can assist with primary prevention of Pb poisoning by conducting geochemical mapping of urban environments. Environmental management of the city cannot take place unless citizens, health care professionals, and politicians are informed about the extent of Pb contamination and remedies for primary prevention. Environmental Pb management has been tested in trials in Minnesota and British Columbia. The pilot project conducted in the Twin Cities of Minneapolis and St. Paul, Minnesota evaluated the hypothesis that soil is a major reservoir for Pb exposure of children. There is a seasonal increase of blood lead during summer months and a corresponding decrease during winter months. Children's blood-Pb was measured at the beginning and end of the project. Two communities were selected where soils contained 500 to 1000 mg Pb /kg. In the target community interventions were taken to reduce children's exposure to bare soil including: establishing a plant cover, topping with wood chips, placing paving stones on barren parking areas, and providing sandboxes with clean sand. The control community children only had blood-Pb measurements at the beginning and end of the project, without interventions. At the end of the summer, the children in the target community did not experience the expected seasonal blood-Pb increase, while children living in the control community experienced a substantial summertime blood-Pb increase (Mielke et al. 1992).

Trail, British Columbia, the site of the largest Pb-Zn smelter in North America, developed an extensive environmental Pb management project. Many neighborhoods in Trail contain 500–1000 mg Pb/kg soil, and similar in quantity to inner city communities of Baltimore, the Twin Cities, and New Orleans. In the 1980s, many Trail children exhibited excessive exposure to Pb. The Trail Pb prevention program advanced the Minnesota pilot project model by using the "stakeholders" approach to solve their community Pb problem (Hilts 1996). Despite initial hesitation, a primary Pb prevention program was instituted. In addition to measures undertaken in Minnesota, extensive efforts were focused on educating parents and child-care providers about washing hands, establishing clean outdoor play areas, and reducing the smelter Pb emissions. The outcome was a significant reduction in the blood Pb of the children of Trail (Hilts 1996, Hilts et al. 1998). The above findings support Clair Patterson's declaration about Pb (Davidson 1999).

Conclusions

Clair Patterson's statement "that the older urban areas of the United States have been rendered more or less uninhabitable by the millions of tons of poisonous industrial lead residues" has worldwide evidence and support. In Nordic countries and Germany, geology and earth science agencies have begun collecting, analyzing, and mapping the geochemistry of urban environments (Birke and Rauch 1997, Birke and Rauch

2000, Bityukova et al. 2000, Trondheim 2000). Urban geochemical maps provide information to health agencies that can be used for identifying problem areas and developing primary prevention strategies to improve the quality of life. If existing knowledge and skills were systematically applied to managing environmental contamination, Pb poisoning in the U.S. could be substantially reduced. In addition, understanding the anthropogenic processes associated with Pb are instructive in preventing urban contamination by other toxic substances. As illustrated by Pb, the skills of geologists and earth scientists linked with medical knowledge, that is, the "medical geology" interdisciplinary approach, are needed to improve and sustain healthy habitation of the planet.

Acknowledgments

Funding for research was from an ATSDR/MHPF cooperative agreement # U50/ATU398948 to Xavier University.

References

Birke, M. and Rauch, U., 1997, Geochemical investigations in the Berlin metropolitan area: Zeitschift Angewandte Geologie, v. 43, p. 58–65.

Birke, M. and Rauch, U., 2000, Urban geochemistry: investigations in the Berlin metropolitan area: Environmental Geochemistry and Health, v. 22(3), p. 233–248.

Bityukova, L., Shogenova, A., and Birke, M., 2000, Urban geochemistry: a study of element distributions in the soils of Tallinn (Estonia): Environmental Geochemistry and Health, v. 22(2), p. 173–193.

Bourgoin, B.P., Evans D.R., Cornett, J.R., Lingard, S.M., and Quattrone, A.J., 1993, Lead content in 70 brands of dietary calcium supplements: American Journal of Public Health v. 83 p. 1155–1160.

Brody, D.J., Pirkle, J.L., Kramer, R.A., Flegal, K.M., Matte, T.D., Gunter, E.W., and Paschal D.C., 1994, Blood lead levels in the US population: Phase 1 of the third national health and nutritional examination survey, NHANES III 1988 to 1991: Journal of the American Medical Association v 272(4) p. 277–283.

Boyd, R., Anderson, D., Crosby, E., Cunningham, L., Frymier, J., Gansneder, B., and Roaden, A., 1999, Are Ohio's urban youth at risk?: urban and suburban schools in the Buckeye state: Bloomington, Indiana, Phi Delta Kappa International, p. 257.

CDC, 1991, Preventing Lead Poisoning in Young Children: Atlanta GA: Centers for Disease Control.

Davidson, C.I. (editor), 1999, Clean hands: Clair Patterson's crusade against environmental lead contamination: Commack, NY, Nova Science Publishers, Inc., 162 p.

Dietert, R.R., Etzel, R.A., Chen, D., Halonen, M., Hollady, S.C., Jarabek, A.M., Landreth, K., Peden, D.B., Pinkerton, K., Smialowicz, R.J., and Zoetis, T., 2000, Workshop to identify critical windows of exposure for children's health: immune and respiratory systems work group summary: Environmental Health Perspectives, v. 108(suppl 3), p. 483–490.

Hilts, S.R., 1996, A cooperative approach to risk and other communities management in an active lead/zinc smelter community: Environmental Geochemistry and Health, v. 18, p. 17–24.

Hilts, S.R., Bock, S.E., Oke, T.L., Yates, C.L., and Copes, R.A., 1998, Effect of interventions on children's blood lead levels: Environmental Health Perspectives, v. 106 (2), p. 79–83.

Mielke, H.W., 1993, Lead dust contaminated U.S. cities: Comparison of Louisiana and Minnesota: Applied Geochemistry v. Supplement Issue 2, p. 257–261.

Mielke, H.W., 1994, Lead in New Orleans soils: New images of an urban environment: Environmental Geochemistry and Health, v. 16(3/4), p. 123–128.

Mielke, H.W., Adams, J.E., Huff, B., Pepersack, J., Reagan,P.L., Stoppel, D., and Mielke, P.W. Jr., 1992, Dust control as a means of reducing inner–city childhood Pb exposure: Trace Substances and Environmental Health, v. 25, p. 121–128.

Mielke, H.W., Adams, J.L., Reagan, P.L., and Mielke, P.W., 1989, Soil-dust lead and childhood lead exposure as a function of city size and community traffic flow: The case for lead abatement in Minnesota: In, Lead in Soil (Davies, B.E. and Wixson, B.G., eds.): Environmental Geochemistry and Health, Supplement 9, 253–27.

Mielke, H.W., Anderson, J.C., Berry, K.J., Mielke, P.W., and Chaney, R.L., 1983, Lead concentrations in inner city soils as a factor in the child lead problem: American Journal of Public Health, v. 73, p. 1366–1369.

Mielke, H.W., Blake, B., Burroughs, S., and Hassinger, N., 1984, Urban lead levels in Minneapolis: The case of the Hmong children: Environmental Research, v. 34, p. 64–76.

Mielke, H.W., Burroughs, S., Wade, S., Yarrow, T., and Mielke, P.W., 1984/85, Urban lead in Minnesota: Soil transects of four cities: Minnesota Academy of Science, v. 50(1), p. 19–24.

Mielke, H.W., Dugas, D., Mielke, P.W., Smith, K.S., Smith, S.L., and Gonzales, C.R., 1997, Associations between soil lead and childhood blood lead in urban New Orleans and rural Lafourche Parishes of Louisiana U.S.: Environmental Health Perspectives, v. 105, p. 950–954.

Mielke, H.W., Gonazales, C.R., Smith, C.R., and Mielke, P.W., 1999, The urban environment and children's health: soils as an integrator of lead, zinc, and cadmium in New Orleans, Louisiana, U.S.: Environmental Research, v. 81, p. 117–129.

Mielke, H.W., Gonzales, C.R., Smith, M.K., and Mielke, P.W., 2000, Quantities and associations of lead, zinc, cadmium, manganese, chromium, nickel, vanadium, and copper in fresh Mississippi delta alluvium and New Orleans alluvial soils: The Science of the Total Environment, v. 246, p. 249–259.

Mielke, H. W., Powell, E., Shah, A., Gonzales, C., and Mielke, P. W., 2001, Multiple metal contamination from old house paints: consequences of power sanding and paint scraping in New Orleans, Environmental Health Perspectives, (accepted for publication March 2001)

Mielke, H.W. and Reagan, P.L., 1998, Soil is an important pathway of human lead exposure: Environmental Health Perspectives, v. 106, p. 217–229.

Mushak, P., Davis, J.M., Crocetti, A.F., and Grant, L.D., 1989, Prenatal and postnatal effects of low-level lead exposure: integrated summary of a report to the U.S. Congress on childhood lead poisoning: Environmental Research, v. 50, p. 11–36.

NAS, 1980, Lead in the human environment: Committee on lead in the human environment, Washington DC: National Academy of Sciences, p. 265–349.

Needleman, H.L., Riess, J.A., Tobin, M.J., Biesecker, G.E., and Greenhouse, J.B., 1996, Bone lead levels and delinquent behavior: Journal of the American Medical Association, 275(22), p. 1727–1728.

Pirkle, J.L., Brody, D.J., Gunter, E.W., Kramer, R.A., Paschal, D.C., Flegal, K.M., and Matte, T.D., 1994, The decline in blood lead levels in the United States: The national health and nutritional examination surveys NHANES: Journal of the American Medical Association, v. 272(4), p. 284–291.

Trondheim, 2000, Jordforurensning I Byer, Oppsummering fra et arbeidsseminar pa Norges geologiske undersokelse I Trondheim, 3. Or 4. April, p. 1–6.

USEPA, 2001, Lead; identification of dangerous levels of lead; final rule: Federal Register, Part III, Environmental Protection Agency, 40 CFR Part 745, (Friday, January 5, 2001), p. 1205–1240.

Viverette, L., Mielke, H.W., Brisco, M., Dixon, A., Schaefer, J., and Pierre, K., 1996, Environmental health in minority and other underserved populations: benign methods for identifying lead hazards at day care centres of New Orleans: Environmental Geochemistry and Health, v. 18, p. 41–45.

Part III

Identifying the Hazards

Geogenic Hazards

A recurring theme throughout this book is the need for government involvement and support in research and monitoring activities, although in view of past experiences in many countries, this seems still to be a pious hope. However, there are some good financial arguments in favor of bringing geological expertise and data to assist the efforts on health locally and even globally. The following chapters describe geochemical approaches and methods that could, if part of regular government services, help to foster preventive medicine and avert human suffering.

Plant et al.* describe the Global Geochemical Baseline Project sponsored by the International Union of Geological Sciences (IUGS), United Nations Educational, Scientific, and Cultural Organization (UNESCO), and several other international scientific associations. The project, now completed for Britain and China, has resulted in a series of maps on which are plotted the present distribution of naturally occurring metals and other elements of interest. The maps, the result of standardized analyses of surface water and stream sediments, record the total amount of the elements. These quantitative data show variations related to the different sources, including soils and rocks, and to the effects of agriculture and industry. The amounts present can be compared with those specified as safe for human health in national regulations or in the international standards set by the World Health Organization (WHO). This kind of detailed chemical information would enable local authorities to be more proactive in public health matters — for example, to warn against potentially unhealthy situations arising from contaminated waters (see **Neal**). In addition to potentially toxic elements such as As, Pb, or Hg, the maps of other elements (e.g. Na, K, P) that show deficiencies could spark programs to add soil supplements to increase agricultural yields. Geochemical surveys could also be expanded to include hazardous organic compounds found in pesticides, or the various industrial chemicals (see **Plant and Davis**). Clearly, immense efforts and funds would be needed to achieve more global coverage, and the dynamics of earth systems require that regular monitoring should be part of the project. The integration of geologic techniques and expertise, especially in obtaining baseline information for areas slated for reclamation or development, would be prudent in order to prevent potentially harmful health effects.

The importance of a more widespread geochemical mapping program is seen in the review by **Davies** of health problems in eastern and southern Africa. These range from deficiencies in I and excesses in radon (Rn), As, Hg, and F, to possible links between heart disease and high Ce in ingested soils. There are also intriguing suggestions of connections

* *References to chapters in this volume are in bold.*

between volcanic soils and Kaposi's sarcoma and podoconiosis. Though the focus by **Davies** is mainly on humans, he also points to work linking the geochemistry of soils with the health of plant and animal communities. Whether or not the carrying capacity of some parts of the continent has now been exceeded is arguable, but **Davies'** survey shows that there are many public health issues that would benefit by more and better integrated geological and medical information.

Turning now to the national scale, another approach that yields important information is presented by **Selinus**. He describes a regional geochemical monitoring system in Sweden that is focused on aquatic plants and mosses, rather than on the stream sediments traditionally used. The advantage of this system is that it averages out short-term variations and reflects elemental bioavailability integrated over time. He illustrates his results with a map showing Cd distribution in southern Sweden that brings up an interesting question. Do the amounts of Cd determined reflect multiple sources, for example, from atmospheric deposition as well as from the minerals in local soils and rocks? Selinus also describes studies investigating potential links between cardiovascular disease and water hardness, and between childhood diabetes and low zinc in drinking water.

Medical Hazards

The elements we require and consume through food and water cause us to directly depend on the environment. The final group of chapters considers the "environment" within the human body, and some of the internal ways in which external geogenic factors can affect the skeleton, teeth, and other organs. Much has and continues to be written on human physiology by and for medical and dental practitioners and researchers (e.g., Rugg-Gunn and Nunn 1999, Simmons 1990). However, to appreciate how the presence and properties of the one mineral species, hydroxylapatite, essential to health, plays a variety of roles, **Skinner** goes beyond the cell systems that produce vertebrate tissues and structures. This calcium phosphate mineral phase provides a ready source of phosphorus — the essential element involved in all energy transfers within the body. Also, because of the peculiar crystal chemical characteristics of hydroxylapatite, many of the other cations and anions needed in the spectrum of human biochemical pathways become part of the mineral; and because of the dynamic bone system, they are locally available and recycled. The chemical composition of the skeleton is a direct reflection of diet, and the ingestion of low levels of some elements, or an excess, can cause debilitation, disease, and possibly death. **Skinner** illustrates the dynamic interrelationships between biochemistry and mineralogy of the human skeleton with the radioactive element strontium, ^{90}Sr. When the U.S. was testing atomic weapons in southwestern U.S. during the 1950s, the potential for uptake and contamination of this hazardous isotope through daily ingested foods was investigated using geochemical techniques on children's deciduous teeth. The cooperative geologic-dental investigations attested not only to the role of the apatite mineral, but also to the processes acting to limit bioavailability of such a hazard.

Ceruti et al. describe skeletal disorders in Maputoland, South Africa. They report the first results of a study of the possible relationship between soils and endemic disease (osteoarthritis and dwarfism), and they focus on possible pathways via the staple crops of the area. Potentially harmful elemental deficiencies are identified relative to the soil nutrient levels considered critical for maize growth. Understanding this and other endemic diseases requires careful study of the links between background chemistry, food crops, and diet.

There can be several reasons for improper skeletal development, some genetic, some biochemical. If these studies can identify a geochemical trigger, nutritional supplements could be a reasonable and appropriate answer. This study is a reminder that there are inherent limitations in some environments inhabited by humans. Even here, however, careful attention to the geological and geochemical background, as determined in the geochemical mapping strategies outlined by **Selinus** and **Plant et al.**, can help to improve the health and well-being of people, and perhaps in a relatively inexpensive manner.

Another health topic that may have a relation to the environment concerns heart attacks and the cardiovascular system (Ross 1997, Yu 1974). In the past, atherosclerotic plaques that calcify to become arteriosclerotic have been ascribed to the "hardness" of domestic water, dependent on the concentrations of Mg and Ca. Epidemiological studies seeking to link disease with these simple chemical differences have always been inconclusive. Other variables, such as dietary preferences of individuals or whole populations, need to be included when evaluating cause and effect. Some simple and very preliminary experimentation on dissolution of mineralized substances in human arteries using chemical methods is presented by **Pawlikowski**. Though he shows laboratory success, clinical adoption is a long way off, and not just because of expense; the seemingly straightforward methodology has enormous potential side effects in humans. However, a few physicians in the U.S. have practices in which arterial cleansing, or chelation therapy, is an elected form of treatment for such disorders and they have willing patients (Sinatra 1996, Walker 1990). Most researchers and clinicians do not recommend such flow-through procedures, although angioplasty and the insertion of stents has become a regular mode of treatment for advanced cardiovascular blockages (Sousa et al. 2001).

Tatu et al. report their ongoing studies of another disease, which manifests itself in a strange kidney ailment endemic to several regions in the Balkan countries. A link to the local low-rank Pliocene lignites is suspected, but not yet proven. The research requires detective work, involving organic geochemists and medical professionals to isolate and identify the potentially offending molecular species. The puzzle is how toxic components in the coal become bioavailable, but the anticipation is that once this is known, it will not only help the local population, but may also contribute to a better global understanding of diseases of the urinary tract.

References

Ross, Russell (1997) Atherosclerosis and inflammatory disease. New England Journal of Medicine, v. 340, p.115–126.

Rugg-Gunn, A.J. and Nunn, June H. (1999) Nutrition, diet and oral health. Oxford University Press. New York. 198 p.

Simmons, David J. (1990) Nutrition and Bone Development. Oxford University Press, New York. 383 p.

Sinatra, Stephen (1996) A cardiologist's prescription for optimum health. Lincoln-Bradley Pub., New York

Sousa, J. Eduardo, Costa, Marco, A. et al. (2001) Lack of neointimal proliferation after sironlimus coated stents in human coronary arteries: a quantitative coronary angiography and three dimensional intravascular ultrasound study. Circulation, v.103(2), p.192–195.

Walker, Morton (1990) The Chelation Way, Avery Publishing Group, Inc., Garden City Park, NY.

Yu, P. (1974) Review of the calcification processes in atherosclerosis. p. 403–425 in Wagner, W.D. and T.B. Clarkson (Eds) Arterial mesenchyme and atherosclerosis. Plenum, New York.

20

Environmental Geochemistry on a Global Scale

JANE PLANT
British Geological Survey, Keyworth, Nottingham, United Kingdom

DAVID SMITH
U.S. Geological Survey, Denver, Colorado

BARRY SMITH AND SHAUN REEDER
British Geological Survey, Keyworth, Nottingham, United Kingdom

Recent population growth and economic development are extending the problems associated with land degradation, pollution, urbanization, and the effects of climate change over large areas of the earth's surface, giving increasing cause for concern about the state of the environment. Many problems are most acute in tropical, equatorial, and desert regions where the surface environment is particularly fragile because of its long history of intense chemical weathering over geological timescales.

The speed and scale of the impact of human activities are now so great that, according to some authors, for example, McMichael (1993), there is the threat of global ecological disruption. Concern that human activities are unsustainable has led to the report of the World Commission on Environment and Development *Our Common Future* (Barnaby 1987) and the establishment of a United Nations Commission on Sustainable Development responsible for carrying out Agenda 21, the action plan of the 1992 Earth Summit in Rio de Janeiro, Brazil.

Considerable research into the global environment is now being undertaken, especially into issues such as climate change, biodiversity, and water quality. Relatively little work has been carried out on the sustainability of the Earth's land surface and its life support systems, however, other than on an ad-hoc basis in response to problems such as mercury poisoning related to artisanal gold mining in Amazonia or arsenic poisoning as a result of water supply problems in Bangladesh (Smedley 1999). This chapter proposes a more strategic approach to understanding the distribution and behavior of chemicals in the environment based on the preparation of a global geochemical baseline to help to sustain the Earth's land surface based on the systematic knowledge of its geochemistry.

Geochemical data contain information directly relevant to economic and environmental decisions involving mineral exploration, extraction, and processing; manufacturing industries; agriculture and forestry; many aspects of human and animal health; waste disposal; and land-use planning. A database showing the spatial variations in the abundance of chemical elements over the Earth's surface is, therefore, a key step in embracing all aspects of environmental geochemistry. Although environmental problems do not respect political boundaries, data from one part of the world may have important implications elsewhere. To achieve a common database with permanent value, it is necessary to adopt standardized procedures for every step of the acquisition process.

Essential and Potentially Harmful Chemicals

Three groups of chemicals are of particular concern in relation to the Earth's life support system: essential and potentially harmful chemical elements; radioactive substances; and persistent organic pollutants.

In the case of chemical elements, two main groups are of particular importance for health: those that are essential to animal life and those that are potentially harmful. Elements essential to animal life include K, Na, Ca, P, Cl, S, and N as well as the first row transition elements Fe, Mn, Ni, Cu, vanadium (V), Zn, Co and Cr, and Mo, tin (Sn), Se, and I. Some of the disorders associated with deficiency of these elements are given in Mertz (1986). Boron has not yet been shown to be necessary for animals, although it is essential for higher plants. By contrast, potentially harmful elements (PHEs) known to have adverse physiological significance at relatively low levels include As, silver (Ag), beryllium (Be), Cd, Pb, Hg, U and some of its daughter products, and possibly the rare earth elements Ce and gadolinium (Gd). Aluminium (Al) can also have adverse physiological effects in trace amounts in animals and plants (Sposito 1989) and has been implicated in some types of neurological diseases in humans (Harrington et al. 1994). All trace elements are toxic if ingested or inhaled at sufficiently high levels for long enough periods of time. Se, F, and Mo are examples of elements that show a relatively narrow concentration range (of the order of a few $\mu g\ g^{-1}$) between deficiency and excess (toxic) levels.

Radionuclides can also be considered in two groups: those that are naturally occurring, and those produced deliberately or accidentally by nuclear reactions as a consequence of human activities. Naturally occurring radionuclides include ^{40}K, ^{235}, ^{238}U, and ^{232}Th, and their daughter products. Man-made radionuclides potentially hazardous to health include ^{137}Cs, ^{95}Zr, and ^{131}I, which are chemically indistinguishable from non-radioactive isotopes of the same elements normally present in the environment.

In the case of synthetic organic (carbon-based) chemicals, most are manufactured by the worldwide chemical industry that produced 400 million tons of chemicals in 1995. According to the European Environment Agency, the exact number of marketed chemicals is unknown; the current estimate varies from 20,000 to 70,000 (Teknologi-Rådet 1997). Little is known about the toxicity of about 75% of these chemicals (Environmental Defence Fund 1997, Natural Research Council 1984). In addition to marketed chemicals, chemical by-products formed by processes such as energy production can also have impacts on the environment.

The organic chemicals that are of most concern are those that are persistent and travel long distances, in some cases on a global scale, especially those that bioaccumulate up the food chain. Some persistent organic pollutants (POPs) are found in the Arctic thousands of miles from major primary sources. Some health effects of chemicals in the environment are listed in Table 20.1.

The link between chemicals and health effect varies from well-known causal relationships such as that between benzene and leukemia, to suggestive associations such as that between pesticides and chemical sensitivity.

Generating a Geochemistry Database

Geochemistry can be used to identify, map, and monitor not only the total amount of substances in soil, dust, or water, but also the amount of a substance that is bioavailable and hence most significant in terms of likely human health effects. For example, Al, which is the most common metal in the Earth's surface layer, occurs in both inert and bioavailable forms, and its potential toxicity depends on its chemical form or speciation. It can occur in a large number of dissolved aqueous species, especially in conditions of low or high pH, or in association colloids with organic carbon or silica that limit its bioavailability and, therefore, toxicity. The toxicity of As and Sb also depends on their chemical form. They are most toxic in the M^{-3} gaseous state with decreasing toxicity in the sequence M^{3+} > M^{5+} > methyl As/Sb (Abernathy 1993, Chen et al. 1994).

Table 20.1: Potential effects on human health of chemicals in the environment (after EEA 1997)

Health Effect [a]	*Sensitive Group*	*Examples of Associated Chemicals*	
Cancer	All	Asbestiform minerals Polycyclic aromatic hydrocarbons (PAHs) Benzene Some metals Some pesticides Some solvents Natural toxins	
Cardiovascular diseases	Especially elderly	Carbon monoxide Arsenic Lead Cadmium	Cobalt Calcium Magnesium
Respiratory diseases	Children, especially asthmatics	Inhalable particles Sulphur dioxide Nitrogen dioxide Ozone	Hydrocarbons Some solvents Terpenes
Allergies and hypersensitivities	All, especially children	Particles Ozone	Nickel Chromium
Reproduction	Adults of reproductive age	Polychlorinated biphenyls (PCBs) DDT Phthalates Other endocrine disrupters	
Developmental	Fetuses, children	Lead Mercury Other endocrine disrupters	
Nervous system disorders	Fetuses, children	PCBs Methyl mercury Lead Organophosphates, including pesticides	Aluminium Organic solvents Manganese

(a) The link between chemicals and health effect varies from well-known causal relationships, such as that between benzene and leukemia, to suggestive associations such as that between pesticides and chemical sensitivity.

As concern about the environment has increased, more emphasis has been placed on studies concerned with issues such as the quality of water resources; waste and pollution related to historical mining, industrialization, and urbanization; and, in developing countries, soil degradation and desertification as a result of salination and deforestation. Such studies make an important contribution to applied environmental research, but are limited by their tendency to investigate specific media. Groundwater studies in particular tend to be carried out in isolation from those of surface water, stream sediment, or soil. There is also a tendency to study one or two chemical elements in detail — an approach that fails to provide a complete picture of interaction between chemical species in the surface environment as a whole. Methods of sampling, analysis, and data interpretation often vary greatly. Moreover, such studies are generally published in the journals of learned societies and are communicated mainly to other experts in the same field with relatively little impact on society as a whole.

In most developed countries, fully integrated digital multimedia, multi-element geochemical data are becoming available as a basis for both economic and environmental studies. When linked to process and modeling studies, such data provide a powerful means of documenting environmental or natural resource problems, as well as providing solutions and subsequently assessing their effectiveness (Plant et al. 2000). Developing countries are increasingly recognizing that geochemical mapping should be given a high priority within their national programs (Reedman et al. 1996).

Some of the applications of data obtained from geochemical mapping include:

- Delineation of areas with mineral resource potential.

- Identification of contaminated land.

- Studies of water quality.

- Studies of the environmental impact of agriculture and forestry.

- Assessment of acid drainage potential and other contamination from mines.

Table 20.2: Summary of principal recommendations for preparing a global geochemical database (after Darnley et al. 1995)

1. Commonly available representative sample media, collected in a standardized manner.

2. Continuity of data across different types of landscape.

3. Adequate quantities of the designed sample media for future reference and research requirements.

4. Analytical data for all elements of environmental or economic significance.

5. Lowest possible detection limits for all elements.

6. Determination of the total amount of each element present.

7. Tight quality control at every stage of the process.

Extending modern systematic geochemical databases such as those prepared by the BGS and USGS over the Earth's land surface clearly provides one of the most effective methods of addressing environmental concerns worldwide. However, available data are incomplete and inconsistent between different countries (Darnley et al. 1995). Many older datasets that were collected mainly for prospecting purposes do not meet the basic requirement for establishing national environmental baselines. Moreover, although there are exceptions, different national and international agencies have carried out different aspects of geochemical surveys using different methodologies. Hence, geological surveys have traditionally provided chemical data on rocks and stream sediments; soil surveys data on soils; and hydrological survey organizations data on the chemistry of ground and surface water. It is difficult to interact such datasets using GIS or to use them as a basis for studying processes operating in the Earth's surface environment as a whole.

The Global Geochemical Baseline Program

The concept of internationally standardized geochemical mapping procedures originated in the 1970s under the auspices of the International Atomic Energy Agency, specifically for uranium. Successful application of the concept led, in 1984, to discussions on the need to have standardized data for all elements. In 1988, formal support was obtained from the IUGS/ UNESCO International Geological Correlation Program (IGCP) for what has become the Global Geochemical Baseline Project (Darnley et al. 1995). Since then, a worldwide network of professional geochemists from over 120 countries has been established. A summary of the principal recommendations for preparing the global geochemistry database is given in Table 20.2.

Initially, a comprehensive review of sampling, analytical, and data processing methods in use was carried out and a framework was developed for standardizing methods for geochemical mapping at the global-regional scale. The recommended sampling, analytical, and data processing methods are described in Darnley et al. (1995); a more detailed sampling manual has been prepared by Salminen et al. (1998). All this information is also available through web sites (Table 20.3).

In order to begin systematic international geochemical mapping, a primary geochemical reference network (GRN) comparable to a geodetic grid has been established, a copy of which is available through the web (Table 20.3). The samples collected for the GRN also serve as standard reference materials. In the case of radionuclides, airborne gamma ray surveys provide the most effective method of mapping and standard procedures have been recommended (Darnley et al. 1995) based on those of the IAEA.

The GRN of geochemical samples has been completed for China and parts of Russia and work is on schedule to complete the work over 27 European countries (Plant et al. 1997, Salminen et al. 1998) by the end of 2002. Australia is carrying out an airborne gamma ray survey of the whole country as a component of the project. A considerable amount of work on developing a geochemical baseline is also being carried out in southern Africa, Colombia, Brazil, South Korea, and many other countries. Geological survey organizations preparing geochemical maps in developing countries are also increasingly using the standard methods, although much more financial assistance is required to extend the network especially over such countries. As indicated by Darnley et al. (1995), the total amount of money required for such an important study of the Earth is no greater than the launch preparation costs for one space shuttle mission.

Conclusions and Recommendations

Preparation of a systematic multimedia (surface and groundwater, soil and stream sediment) multi-determinand global geochemical database would make a major contribution to understanding and hence improving environmental quality worldwide. In the longer term, the aim should be to include POPs. Such a program could involve all countries with the communication of methods, data, and information over the world wide web. This would provide the best method of technology transfer, and for documenting and communicating information on the environment of particular regions and countries in a global context as a basis for sustainability. It would also provide a sound basis for targeting scarce resources and developing strategies for remediation and amelioration. A global database would also provide a baseline against which future change could be measured.

Table 20.3: List of web sites

Organization	Acronym	Internet or Web Site Address
British Geological Survey	BGS	http://www.bgs.ac.uk/iugs
U.S. Geological Survey	USGS	http://www.usgs.gov
Geological Survey of Finland	GSF	http://www.gsf.fi/foregs
International Union of Geological Sciences	IUGS	http://www.iugs.orgiugs/science/sci_wggb.html
International Association of Geochemistry and Cosmochemistry	IAGC	http://www.cevl.msu.edu/~long/IAGC/html
International Geological Correlation Programme	IGCP	http://www.unesco.org/science/earthsciences/igcp
United Nations Educational, Scientific and Cultural Organization	UNESCO	http://www.unesco.org

References

Abernathy, C.O. 1993. Draft drinking water criteria document on arsenic. US-EPA Science Advisory Board report, contract 68-C8-0033.

Barnaby, F. 1987. Our Common Future – The Brundtland-Commission Report. Ambio, v.16, p.217–218.

Chen, S.L., Dzeng, S.R. & Yang, M.H. 1994. Arsenic species in the Blackfood disease area, Taiwan. Environmental Science & Technology, v.28, p.877–881.

Darnley, A.G., Björklund, A., Bølviken, B. & 8 others 1995. A Global Geochemical Database for Environmental and Resource Management. Recommendations for International Geochemical Mapping. Earth Science Report 19, UNESCO Publishing, Paris.

EEA — European Environment Agency. 1997. Chemicals in the European Environment: LowDoses, High Stakes? The European Environment Agency and United Nations Environment Programme Annual Message 2 on the State of Europe's Environment. UNEP/ROE/97/16.

Environmental Defence Fund. 1997. Toxic Ignorance: The Continuing Absence of Basic Health Testing for Top Selling chemicals in the US, Environmental Defense Fund, Washington.

Harrington, C.R., Wischik, C.M., McArthur, F.K., Taylor, G.A., Edwardson, J.A. & Candy, J.M. 1994. Alzheimer's-disease-like changes in tau protein processing: association with aluminium accumulation in brains of renal dialysis patients. Lancet, v.343, p.993–997.

McMichael, A.J. 1993. Planetary Overload. Global environmental change and the health of the human species. Cambridge University Press.

Mertz, W. 1986. Trace elements in human and animal nutrition. US Dept of Agriculture. Academic, Orlando.

Natural Research Council. 1984. Toxicity Testing, National Academy Press, Washington.

Plant, J.A., Klaver, G., Locutura, J., Salminen, R., Vrana, K., Fordyce, F.M. 1997. The Forum of European Geological Surveys Geochemistry Task Group: Geochemical inventory. Journal of Geochemical Exploration, v.59, p.123–146.

Plant, J.A., Smith, D., Smith, B & Williams, L. 2000. Environmental geochemistry at the global scale. Journal of the Geological Society, London, v.157, p.837–849.

Reedman, A.J., Calow, R.C. & Mortimer, C. 1996. Geological Surveys in developing countries: strategie: for assistance. British Geological Survey Technical Report WC/96/20.

Salminen, R. et al. 1998. FOREGS Geochemical Mapping, Field Manual. Geological Survey of Finland, Guide 47.

Smedley, P.L. 1999. Environmental Arsenic Exposure: Health Risk and Geochemical Solutions. BGS Technical Report for DFID.

Sposito, G. 1989. The Environmental Chemistry of Aluminium. CRC, Boca Raton, Florida.

Teknologi-Rådet 1997. The Non-assessed Chemicals in EU. Presentations from the conference 30 October 1996. Danish Board of Technology, Copenhagen.

21

Biogeochemical Monitoring in Medical Geology

O. SELINUS

Geological Survey of Sweden, Uppsala, Sweden

Biogeochemical Mapping

How can we determine the distribution of metals and other elements in our environment? The Geological Survey of Sweden started an innovative monitoring of metals in a monitoring/mapping program in 1980. Before 1980, traditional inorganic stream sediments were used, a method still employed all over the world, but not really suitable for medical work. A new method is used, whereby metal concentrations are determined in organic material consisting of aquatic mosses and roots of aquatic higher plants. These are barrier-free with respect to trace metal uptake and reflect the metal concentrations in stream water (Brundin 1972, 1988, Kabata-Pendias,1992, Selinus 1989). Aerial parts of many plant species do not generally respond to increasing metal concentrations in the growth medium because of physiological barriers between roots and above-ground parts of plants. These barriers protect them from uptake of toxic levels of metals into the vital reproductive organs. The roots and mosses, however, respond closely to chemical variations in background levels related to different bedrock types in addition to effects of pollution. The biogeochemical samples provide information on the time-related bioavailable metal contents in aquatic plants and in the environment. One great advantage of using biogeochemical samples instead of water samples is also that the biogeochemical samples provide integrated information of the metal contents in the water for a period of some years. Water samples suffer from seasonal and annual variations depending on, for example, precipitation.

The mapping program now covers about 65% of the land area of Sweden (40,000 sample sites, one sample every 6 km^2), where about 80% of the population of Sweden is living. This means that there is now available an extensive analytical data base for use in environmental and medical research (Freden 1994).

Cadmium in Southern Sweden

One example of the use of biogeochemical monitoring concerns high cadmium contents in Sweden. In noncontaminated, noncultivated soils, Cd concentration is largely governed by the amount of Cd in the parent material (Thornton 1986). If the substrate concentration is higher than in background concentrations, Cd is readily taken up by roots and is distributed

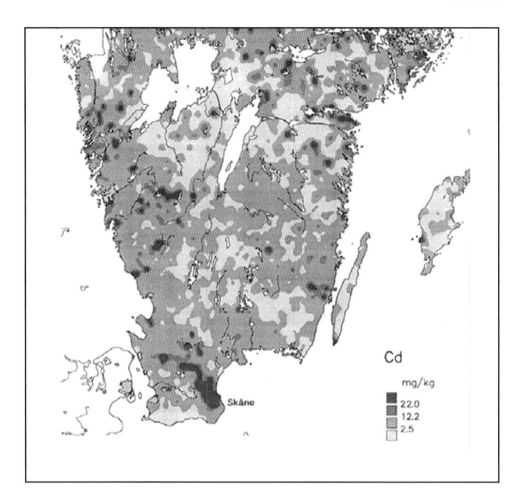

Figure 21.1: Cadmium (Cd) in biogeochemical samples in southern Sweden.

throughout the plants. The amount of uptake is influenced by soil factors such as pH, cation exchange capacity, redox potential, phosphatic fertilization, organic matter, other metals, and other factors. In general, there is a positive, almost linear correlation between the different Cd concentrations in the substrate and the resulting Cd concentration in the plant tissues (Selinus 1983, 1988).

Cadmium levels in biogeochemical samples (roots and aquatic mosses) from southern Sweden are shown in Figure 21.1. The contents are enhanced in the southernmost counties (Skåne) of Sweden and along the west coast of Sweden. The latter distribution is derived mainly from transboundary atmospheric transport and deposition of anthropogenic origin. The contents of Cd in the southernmost part are, however, much higher, and the highest levels so far detected since 1982 are located in this region. This region is a densely populated farming region from which growing crops

are distributed to the rest of Sweden. Samples of autumn wheat have been taken in certain geographical regions. The results showed that samples from Skåne had an average of 73 µg Cd/kg dry weight, with several areas exceeding 100 µg Cd/kg dry weight. In comparison, an area in central Sweden yielded on analysis only 29 µg Cd/kg dry weight on average in autumn wheat. The Cd contents in wheat from Skåne are therefore a matter for concern. In this region drinking water is taken from many wells. In those that have been analyzed for Cd, the results show an almost identical distribution of high Cd contents as depicted in the biochemical map. For drinking water, the WHO has set a limit of 5 µg Cd/l. In comparison, the wells in Skåne, within the region with high Cd burden, have levels of max 400 µg Cd/l.

Several factors may interact with each other: for example, deposition of airborne Cd, as well as acid rain from Eastern, Western, and Central Europe. Cadmium may also originate from phosphate fertilizers used in agriculture. However recent studies have shown that the high contents of Cd are probably derived from sandstones with disseminated Cd. Therefore, we have a reason to believe that we have a connection between geology, acidification, and possible health effects caused by cadmium.

Use of Biogeochemistry in Medical Geology

The inverse relationship between cardiovascular disease and water hardness was first reported in the U.S. in 1956 and from Japan in 1957. A significant inverse relationship has been found between water hardness and total cardiovascular mortality. A higher sudden death rate in soft water areas compared to hard water areas are also reported. In the WHO myocardial infarction registry network, all cases of myocardial infarction were registered in a standardized way in 15 WHO countries in Europe. Higher rates of ischaemic heart disease (IHD) were found in towns served by soft water than in towns with hard water. In Sweden it has shown that water hardness (Ca + Mg and other minor constituents) and the sulphate and bicarbonate concentrations of the drinking water were inversely related to IHD as well as stroke mortality. The variation in the drinking water composition in these areas reflects the geological variation in the region as demonstrated by the biogeochemical sampling program (Nerbrand 1992).

Childhood diabetes is almost exclusively of the autoimmune insulin-dependent type (type1). The genetic prerequisites are clearly not sufficient causes of the disease. Descriptive studies from population-based incidence registers from the countries in Scandinavia have shown a significant within-country geographical variability in insulin-dependent diabetes incidence rates that cannot be explained by a slight south-north gradient only. A case control study has been designed comparing cases and controls as to estimates of zinc obtained in biogeochemical samples from areas of residence. A high water concentration of zinc was associated with a significant decrease in risk. This provides evidence that a low groundwater content of zinc that may reflect long-term exposure through drinking water is associated with later development of childhood onset diabetes (Haglund 1996).

Another example of the use of biogeochemistry in medical geology is the work on moose (Frank). The biogeochemical technique and animal monitoring technique by using moose as a monitor for mapping the bioavailabilty of elements in the environment complement each other on different levels in the food chain. The results using biogeochemistry and those using organ tissues from the moose collected during the same period of time appear to be remarkably similar, despite higher trophic level of the latter. The two methods elucidate the usefulness of the techniques in monitoring and detecting metal burdens of regions on toxic levels as well as deficiency of essential elements

(Selinus 1996, 2000).

References

Brundin, N.H. & Nairis, B. 1972. Alternative sample types in regional geochemical prospecting. J Geochem Explor v.1, p,7–46.

Brundin, N.H., Ek, J.I., Selinus, O.C., 1988: Biogeochemical studies of plants from stream banks in northern Sweden. J. Geochem. Explor., v.27, p.iogeochemical mapping programme of Sweden. EUG V, Strasbourg.

Freden, C., (Editor), 1994. Geology. The National Atlas of Sweden.

Haglund, B., Ryckenberg, K., Selinus, O., Dahlqvist, G., 1996. Evidence of a relationship between childhood-onset type 1 diabetes and low groundwater concentration of Zinc. Diabetes Care, Vol 19, No 8, August 1996.

Kabata-Pendias A. & Pendias, H. 1992. Trace elements in soils and plants. CRC press.

Nerbrand, C., Svärdsudd, K., Ek, J., Tibblin, G., 1992. Cardiovascular mortality and morbidity in seven counties in Sweden in relation to water hardness and geological settings. European heart journal, v.13, p.721–727.

Selinus, O. 1983: Regression analysis applied to interpretation of geochemical data at the Geological Survey of Sweden. In Handbook of exploration geochemistry, part 2, Statistics and data analysis in Geochemical prospecting. Ed R.J. Howarth. Pp 13–19. Elsevier.

Selinus, O., 1988: Biogeochemical mapping of Sweden for geomedical and environmental research. In: Ed:Thornton. Proceedings of the 2:nd symposium on geochemistry and health. Science reviews Ltd, Northwood, U.K.

Selinus, O. 1989. Heavy metals and health — results of the biogeochemical mapping programme of Sweden. EUG V, Strasbourg.

Selinus, O, Frank, A., Galgan, V. 1996. Biogeochemistry and metal biology — An integrated Swedish approach for metal related health effects. In: Environmental geochemistry and Health in Developing Countries. Ed. Appleton, D., Fuge, R., McCall, J. Special Publication. Geological Society of London, no 113, pp 81–89 Chapman and Hall.

Selinus, O., Frank, A., 2000. Medical geology. In Möller, L., (Editor): Environmental medicine. pp 164–183. Joint Industrial Safety Council.

Thornton, I. 1986. Geochemistry of cadmium. In Cadmium in the environment (ed Mislin, H. & Ravera, O). p 8 Birkhauser Verlag.

22

Some Environmental Problems of Geomedical Relevance in East and Southern Africa

T. C. Davies

School of Environmental Studies, Moi University, Eldoret, Kenya

Introduction

Medical geology studies the influence of geo-environmental factors on the geographical distribution of diseases of humans and animals. In the east and southern African subregion, there has been little attention paid to date on the extent to which these factors may be important in disease causation, even though developing countries in general can be shown to hold tremendous promise for specific research in this field. This chapter highlights some problems of geomedical relevance in the subregion and submits that interdisciplinary research among scientists can help provide practical solutions.

Iodine

The iodine deficient regions of east and southern Africa have been identified and the widespread occurrence of goiter and related conditions, collectively referred to as iodine deficiency disorders (IDD), firmly established (e.g., Davies 1994, Jooste et al. 1997).

These are serious and debilitating consequences, particularly for poor populations, as the capacity of children is severely restricted and they become a burden to the family. The reported geographical distribution of endemic goiter in East Africa is shown in Figure 22.1.

Many aid agencies and governments have attempted to solve the problem by increasing dietary intake of iodine via the introduction of iodized salt and iodized oil programs. Despite these interventions, IDD remain a major problem in the subregion. It is likely that IDD are multi-causal diseases involving factors such as trace element deficiencies, goiter-inducing substances in foodstuffs (known as goitrogens), and genetics (Fordyce 2000). However, geochemists have an important role to play in determining the environmental cycling of iodine and its uptake into the food chain if levels of dietary iodine are to be enhanced successfully.

Fluorine

It has now been established that excessive fluorine (mainly in the form of fluoride) is present in parts of the hydrological system of Kenya as well as other countries in the subregion, particularly those that are associated with rift formation (Gaciri and Davies 1993).

Fluoride in minor amounts (around 1.3 ppm) reduces dental decay and enhances the proper development of the bone. A similar level of fluoride intake may also be beneficial to animals. When the amount of fluoride consumed is either too low or much too high, undesirable physiological consequences appear, such as dental caries, mottled staining of the teeth and malformed bone structure in both humans and animals.

Fortunately for Kenya, numerous data now exist on the geochemistry of fluoride in the hydrological system of the country (e.g., Gaciri and Davies 1993,

Figure 22.1: Distribution of reported goiter areas in Eastern Africa from Hanegraaf and McGill (1970) (redrawn).

Nair et al. 1984), the toxicity problem (e.g., Bakshi 1974, Fendall and Grounds 1965) and some improved techniques for fluoride determination (Njenga 1982). It has therefore been possible to put forward practical guidelines for combating the fluoride toxicity problem in the country (Gaciri and Davies 1993).

Endomyocardial Fibrosis

In the late 1980s, medical workers from India noted a strong link between endomyocardial fibrosis (EMF), a fatal coronary heart condition in children throughout the tropics, and enhanced environmental levels of cerium, related to the presence of monazite sands in Kerala province (Smith 1999).

In a collaborative study, Ugandan scientists from Mulago Hospital, the Institute for Child Health, Makerere University, and colleagues from the British Geological Survey have investigated whether similar environmental exposure could account for the occurrence of endemic EMF in Uganda (Masembe et al. 1999). Case-control studies performed at Mulago Hospital, Kampala, indicate a high incidence of EMF

among patients from Mukono and Luwero districts over the past 30 years. Model calculations based on the data collected in this work (Smith et al. 1998) indicate that the most important exposure route is through the ingestion of soil-bound cerium as a result of either the inadvertent oral ingestion of soil or through the habitual eating of soil (geophagy).

Geophagy

Geophagy — the involuntary or sometimes the deliberate eating of earth — is extremely common among traditional African societies and has been recognized since ancient times.

The British Geological Survey has been undertaking research to investigate the potential bioavailability of major and trace elements within soils that are commonly ingested in Uganda (Smith 2000). The objectives of this project are to (a) increase our understanding of the risks and benefits associated with soil ingestion (deliberate and/or inadvertent); and (b) to enable the bioavailability of a particular contaminative source to be accurately taken into account during site specific risk assessment.

The reasons why soils are being deliberately consumed can be difficult to establish, but known causative explanations include the use of soil during famine, as a food detoxifier, as a pharmaceutical, and for neuropsychiatric and psychological (comforting) reasons (Abrahams 2000, and **Abrahams**). It remains a matter of conjecture whether ingestion of soil actually satisfies a nutritional deficiency. Nevertheless, soils do have the potential to supply mineral nutrients, especially iron, where the ingestion of soil can account for a major proportion of the recommended daily intake.

Typical quantities of soil eaten by practicing geophagics in Kenya in the order of 20 g per day have been recorded (Smith 2000). Although the eating of such large quantities of soil increases exposure to essential trace nutrients, it also significantly increases exposure to potentially toxic trace elements and biological pathogens. The former is particularly likely in mineralized areas associated with mineral extraction, or polluted urban environments, where levels of potentially toxic trace elements are high.

Asbestos

Asbestos is a naturally occurring fibrous mineral silicate that is used in a variety of applications in Africa including insulation, pipe lagging, roofing, and brake linings. Asbestos fibers are long and thin, and their aerodynamic characteristics allow them to penetrate far into the lung. In the lungs, asbestos fibers are encased in proteinaceous materials and appear in sputum (Mashalla et al. 1998, see also **Hillerdal**).

The mining of amphibole asbestos is almost completely confined to South Africa. Crocidolite and amosite, locally abundant and mined asbestos minerals, occur in metamorphosed Precambrian sedimentary strata (banded ironstones) in the Transvaal Super-Group. About 75% of the asbestos mineral production is used in the manufacture of asbestos-containing cement pipe.

Despite all available evidence on the occupational hazards related to asbestos exposure, asbestos is still widely used in many African countries where precautionary measures are hardly applied due to high costs of maintaining health monitoring equipment and personnel.

Arsenic in Mine Waters

The exploitation of gold and base-metal deposits can result in acid mine drainage (AMD) generated through inorganic and microbially mediated sulphide oxidation. Further, arsenic contamination of surface drainage and groundwater is documented in many regions of the developing world (Williams 2000), but the overall significance of this hazard remains difficult to quantify due to a paucity of data for many African countries, often with substantial mining activity.

Extreme examples with drainage acidities (below pH 1.0) are relatively rare, but acid mine drainage with a pH of 0.52 was encountered at Iron Duke Mine near Mazowe, Zimbabwe, in February 1994 during an investigation of the environmental geochemistry of mine waters in the Harare, Shamva, and Midlands Greenstone Belts of Zimbabwe. Arsenic values of up to 72 mg l^{-1}, recorded at Iron Duke, constitutes the highest dissolved arsenic concentration published to date for mine waters worldwide (Williams and Smith 2000). The site provides a valuable opportunity for studies of acutely acid mine waters, and may assist in validating and refining models for such systems. The chemical toxicity of the AMD generated at Iron Duke precludes most aquatic life immediately downstream of the waste pile, with the effect being highly localized.

Environmental arsenic exposure is a causal factor in human carcinogenesis and numerous non-cancer health disorders. Chronic exposure symptoms most commonly include hyperkeratosis, hyperpigmentation, skin malignancies, and peripheral arteriosclerosis (Blackfoot Disease), all of which are known in populations consuming water with 100–1000 µg l^{-1} arsenic (**Finkelman et al.**).

Buffering methods to control AMD have, in some instances, shown considerable potential for removing arsenic from mine drainage waters (Breward and Williams 1994). Sequential extraction speciation studies and geochemical modelling methods have demonstrated that arsenic is highly prone to scavenging and precipitation with hydrous ferric oxides across a wide pH range (Breward and Williams 1994).

Mercury Exposure in Small-scale Gold Panning

Increasing health problems associated with mercury contamination, as a result of its use as an amalgam agent by small-scale alluvial gold miners, have been reported in many African countries during the past decade (Ghana, Kenya, Zimbabwe, South Africa, etc). Approximately two tons of mercury is used for each ton of gold recovered. Less efficient operations may use even more. With global gold production by small-scale miners now at a rate of thousands of tons per annum, the design and implementation of appropriate methods for monitoring mercury contamination in areas of alluvial gold mining are urgently required. The growth of independent small-scale gold mining provides employment and generates wealth, but economic gains are frequently offset by health degradation and the destruction of vital industries such as fishing.

In Africa, problems due to mercury and arsenic contamination associated with gold mining are a result of lack of environmental concern (or lack of enforcement of regulations) and, in some cases, inappropriate mining and operating practices. Geochemical

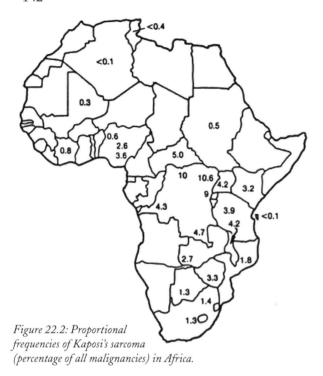

*Figure 22.2: Proportional
frequencies of Kaposi's sarcoma
(percentage of all malignancies) in Africa.*

surveys prior to operation and close monitoring during mining, coupled with good planning and working procedures can minimize contamination, while remediation schemes can also be devised and monitored by the application of sound geochemical science (Breward and Williams 1994).

Kaposi's Sarcoma and Podoconiosis

Kaposi's sarcoma (pronounced *Kawposhi*) is characterized by the growth of multiple vascular nodules on the skin of one or more extremities with occasional involvement of internal organs.

Podoconiosis (non-filarial elephantiasis) is characterized microscopically by neo-volcanic submicron mineral particles (mainly composed of aluminium, silicon, titanium, and iron) stored in the phagolysosomes of macrophages within lymphoid tissues of the lower limbs.

In 1961, at an international conference held at Makerere Medical School in Kampala, Uganda, the high incidence of Kaposi's sarcoma (KS) in the black male population of sub-Saharan Africa was noted. At a second conference in Kampala in 1980, particularly high rates were noted from central, east, and southern African countries, especially the former Zaire (now the Democratic Republic of Congo), Rwanda, Burundi,

Uganda, Malawi, Tanzania, Zambia, Zimbabwe, and Kenya (Fig. 22.2). Incidences decreased toward west and southern Africa (Hutt 1981). Time-space clustering had been observed in the West Nile district of Uganda. The observation that 25 to 30% of individuals with AIDS have KS has given considerable prominence to KS. Thus, KS occurs in sporadic form [as originally described by Kaposi in southern Europeans (cited by England 1961)], in endemic form (as was seen in central African countries prior to the AIDS epidemic), and more recently in epidemic form in association with AIDS. In Africa, all three epidemiological varieties occur. Recent observations indicate that the epidemic form is on the increase in countries such as Uganda and Zambia (Bayley 1984, Lindsay and Thomas 2000).

Many theories of the etiology of KS have been postulated, among which is a geographically linked environmental factor. This nodular condition is believed to arise in the lymphatic endothelium and is associated with chronic lymphoedema. As such, KS bears a superficial resemblance to podoconiosis. The geographical prevalence of both conditions in highland areas of moderate rainfall in proximity to volcanoes suggests a pathogenetic relationship to exposure to volcanic soils. The geographical endemicity of KS in the Congo-Nile watershed (western branch of the East African Rift System) and Nigeria-Cameroon border (Benue Trough and Cameroon Volcanic Line) is particularly striking (Ziegler 1993).

Podoconiosis is also largely endemic in Africa, especially among populations living in areas of heavy exposure to volcanic clay minerals (Price 1990). These iron-oxide-rich kaolinitic soils overlie regions of alkaline basalt associated with volcanism of the major African intercontinental rift systems. The pathogenesis of podoconiosis is consistent with the theory that ultrafine clay minerals from the soil are absorbed through the feet. The resulting chronic lymphatic irritation, inflammation, and collagenesis cause obstruction and lymphoedema of the affected limb. The geographical proximity of endemic KS to areas containing volcanic clay minerals, its lympho-endothelial origin, predilection for the feet and legs, and its prevalence among rural dwelling peasants and cultivators, suggest a common etiology, namely chronic dermal exposure to volcanic clay minerals.

Radiation and Radon Gas

Radon is a colorless, odorless, radioactive gas formed naturally by the radioactive decay of uranium that occurs in all rocks and soils. There is a direct link between the levels of radon generated at the surface and the underlying rocks and soils. Radon generated in shales and granite can reach the surface via faults and fractures.

Variations in radon levels are evident between different parts of the African continent. The principal basis for present concern about radon and its decay products focuses on exposures resulting from industrial processes — primarily the mining and milling of uranium — that increases the accessibility of radon to the outdoor atmosphere or to indoor environments, leading to the incidence of lung cancer. Uranium occurs in a variety of ores, often in association with other minerals such as gold, phosphate, and copper, and may be mined by open-cast, underground, or *in situ* leach methods, depending on the circumstances.

South Africa has Africa's largest identified uranium resources (currently estimated at over 241,000 metric tons), followed by Niger, Namibia, and Gabon. Uranium has also been found in Algeria, Botswana, the Central African Republic, Chad, Egypt, Guinea, Madagascar, Mali, Mauritania, Morocco, Nigeria, Somalia, Tanzania, Togo, Zaire, and Zambia. However, under current economic conditions, it is unlikely that any of these deposits will be exploited in the immediate future.

South Africa produces uranium concentrates as a by-product of gold mining and copper mining, and possesses uranium conversion and enrichment facilities. Uranium is chiefly used as a fuel in nuclear reactors for the production of electricity.

The processes involved in radon gas generation and its movement to the surface and into buildings are complex. These include interactions between rocks, soils, and underground waters. It is important to determine the natural levels of radon produced from rocks and soils as these will provide essential data indicating the potential for high levels of radon entering buildings.

Trace Elements in Soils and Plants — Implications for Wildlife Nutrition

The following case study is an excerpt largely from the conclusions of a paper by Maskall and Thornton (1991) on the Kenyan situation.

Broad variations in the concentrations of some trace and major elements are attributable to differences in soil parent material, but pedogenic and hydrological processes also influence the distribution in soil profiles of trace and major elements, of which the latter (hydrological processes) can affect soil pH and trace and major element uptake into plants and animals, with consequent imbalances in the food chain.

In Lake Nakuru National Park, trace element concentrations in plants varied with the species. Solonetz soils develop on old lake sediments in the presence of sodium cations and carbonate and bicarbonate anions. The geochemistry of these soils nearer to Lake Nakuru appears to greatly increase the availability of molybdenum to the grass *Sporobulus spicatus*.

In general, higher concentrations of copper and cobalt were found in soils developed on basalts than in soils developed on phonolites, trachytes, and volcanic ash. Both impala at Lake Nakuru National Park and black rhino at Solio Wildlife Reserve have a lower blood copper status than animals from other areas. At Lake Nakuru National Park, the low soil copper content and high molybdenum content of some plants probably contribute to the low copper status of impala and may pose a problem for other grazing species.

Application of critical values determined by Maskall (1991) used for domestic ruminants indicates trace element deficiencies in impala at Lake Nakuru National Park and in black rhinoceros at Solio Wildlife Reserve.

Conclusions

Disease seldom has a one-cause, one-effect relationship. The geologist's contribution is to help isolate aspects of the geological environment that may influence the incidence of disease. A unique combination of geochemical factors, traditional practices, cultural norms, and genetics provides an ideal setting for these kinds of study in Africa. But the task is tremendously complex and requires sound scientific inquiry coupled with interdisciplinary research with physicians and other scientists. Although most of the picture is still rather unclear, the possible rewards of this emerging field of medical geology are exciting and may eventually play a significant role in environmental health.

References

Abraham, P.W., 2000. Geophagy by human societies. 'Health and the Geochemical Environment', Abstract Volume, International Working Group on Medical Geology, Uppsala Meeting, 4–6 September, 2000.

Bahshi, A.K., 1974. Dental conditions and dental health. In : L.C. Vogel, A.S. Muller, R.S. Odingo, Z. Onyango and A. de Geus (Eds) : Health and Disease in Kenya, p. 519–522. East African Literature Bureau, Nairobi, Kenya.

Bayley, A.C., 1984. Aggressive Kaposi's sarcoma in Zambia. Lancet, p. 1318.

Breward, N. and Williams, M., 1994. Arsenic and mercury pollution in gold mining. Mining Environmental Management, p. 25–27.

Davies, T.C., 1994. Combating iodine deficiency disorders in Kenya : the need for a multi-disciplinary approach. International Journal of Environmental Health Research, V. 4, p. 236–243.

England, N.W.J., 1961. Histopathological determination of three cases of Kaposi's sarcoma in West Africans; one showing metastasis in an inguinal node. Transactions of the Royal Society of Tropical Medicine and Hygiene, V. 55, p. 301.

Fendall, N.R.E. and Grounds, T.G., 1965. The incidence and epidemiology of disease in Kenya. Part 1. Some diseases of social significance. Journal of Tropical Medicine and Hygiene, Vol. 68, p. 77–84.

Fordyce, F., 2000. Geochemistry and health — why geoscience information is necessary. Geoscience and Development, No. 6, p. 6–8.

Gaciri, S.J. and Davies, T.C., 1993. Occurrence and geochemistry of fluoride in some natural waters of Kenya. Journal of Hydrology, V. 143, p. 395–412.

Hanegraaf, T.A.C. and McGill, P.E., 1970. Prevalence and geographic distribution of endemic goitre in East Africa. East African Medical Journal, V. 47. p. 61.

Hutt, M.S.R., 1981. The epidemiology of Kaposi's sarcoma. In : Olweny, C.L.M., Hutt, M.S.R., Owor, R. (Eds) : Kaposi's Sarcoma, p. 3–8. Basel, Karger.

Jooste, P.L., Weight, M.J. and Kriek, J.A., 1997. Iodine deficiency and endemic goitre in the Langkloof area of South Africa. South African Medical Journal, V. 87, p. 1374–1379.

Lindsay, S.W. and Thomas, C.J., 2000. Mapping and estimating the population at risk from lymphatic filariasis in Africa. Transactions of the Royal Society of Tropical Medicine and Hygiene, V. 94, p. 37–45.

Masembe, V., Mayanya-Kizza, H., Freers, J., Smith, B., Hampton, C., Tiberindwa, J.V., Rutakingirwa, M., Chenery, S., Cook, J., Styles, M., Tomkins, A. and Brown, C., 1999. Low environmental cerium and high magnesium : Are they risk factors for endomyocardial fibrosis in Uganda? Abstract Volume, East and Southern Africa Regional Workshop on Geomedicine, Nairobi, Kenya, p. 13.

Mashalla, Y.J.S., Ntogwisangu, J.H. and Mtinangi, B.L., 1998. Pulmonary function tests in asbestos factory workers in Dar es Salaam. MEDICOM, The African Journal of Hospital Medicine, V. 13, p. 11–14.

Maskall, J. and Thornton, I., 1991. Trace element geochemistry of soils and plants in Kenyan conservation areas and implications for wildlife nutrition. Environmental Geochemistry and Health, V. 13, p. 93–107.

Nair, K.R., Manji, F. and Gitonga, J.N., 1984. The occurrence and distribution of fluoride in groundwaters of Kenya. Challenges in African Hydrology and Water Resources (Proceedings of the Harare Symposium). IAHS Publication, v.144, p. 75–86.

Njenga, L.W., 1982. Determination of fluoride in water and tea using ion selective electrode and colorimetric methods. M. Sc. dissertation, 113 p. University of Nairobi, Nairobi, Kenya.

Price, E.W., 1990. Podoconiosis / non-filarial elephantiasis. Oxford University Press, New York, 131 p.

Smith, B., 1999. Cerium and infantile heart disease. EARTHWORKS, Issue 9, p. 8. DFID Newsletter, Nottingham, U.K.

Smith, B., 2000. Exposure to soil. Geoscience and Development, No. 6, p. 12.

Smith, B., Chenery, S.R.N., Cook, J.M., Styles, M.T., Tiberindwa, J.V., Hampton, C., Freers, J., Rutkinggirwa, M., Sserunjogi, L., Tomkins, A.M. and Brown, C.J., 1998. Geochemical and environmental factors controlling exposure to cerium and magnesium in Uganda. Journal of Geochemical Exploration, Vol. 65, p. 1–15.

Williams, M., 2000. Arsenic in mine waters : an empirical database for the tropics. Geoscience and Development, No. 6, p. 9–12.

Williams, T.M. and Smith, B., 2000. Hydrochemical characterization of acute acid mine drainage at Iron Duke Mine, Mazowe, Zimbabwe. Environmental Geology, Vol. 39, p. 272–278.

Ziegler, J.L., 1993. Endemic Kaposi's sarcoma in Africa is associated with exposure to volcanic clay soils. Abstract Volume, International Conference on Environmental Geochemistry and Health in Developing Countries. The Geological Society, London, 72 p.

23

Geochemistry and Vertebrate Bones

H. Catherine W. Skinner
Department of Geology and Geophysics
Yale University, New Haven, Connecticut

Introduction

The human body has been a focus of attention for thousands of years. The ancients wished to mummify it to assure transposition to the life hereafter. Today we expend a lot of effort and money to forestall the effects of aging of such a complicated machine. The mineralized portion of the body, the skeleton, is the most permanent portion of the body and records the basic size and shape of the individual. Underscoring the wish for properly functioning bones and teeth, we today have a medical 'spare parts' industry that provides substitute knees, hips, or an entire denture. The need to accomplish these implants/transplants successfully has aligned physicians and dentists with cell and molecular biologists, materials scientists, bioengineers, and mineralogists. All wish to create faithful replicas of the mineralized parts, but any substitutes must be in harmony with the internal dynamic biochemical environment, and be part of a viable functioning skeleton.

The general health of every human is a response to the local, regional, and global environment. The interaction noted in the scientific literature almost 50

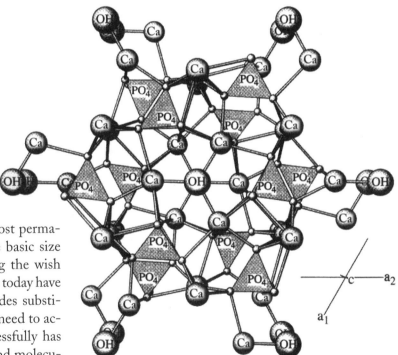

Figure 23.1: *Hydroxylapatite: $Ca_5(PO_4)_3(OH)$*

years ago (Warren 1954) continues (Hopps and Cannon 1972, Ross and Skinner 1994). There is burgeoning interest as many scientists, with ever increasing abilities to detect and measure extremely small amounts of certain elemental species or potentially hazardous substances, focus on the relationships of the environ-

145

ment and health. The roles of the skeletal mineral substance as a participant and faithful recorder in this crossover are outlined here.

Bone Tissues, Cells and Molecular Biology

"Bone" can refer to any one of the more than 200 individual organs with distinctive shapes that make up every human skeleton or to the tissue that is present in each of these organs. The tissues of bone, and teeth, are composites of organic matrix, mineral matter, and specialized cells (Skinner 1987, Albright and Skinner 1987). The cells not only form but also maintain, reconstruct, and repair the bone tissues as required. Cells known as osteoblasts produce a protein and polysaccharide matrix in and on which mineral is deposited. The cells become embedded in this extracellular mineralized matrix but continue to assist and sustain the viability of their products. They have a programmed life cycle and, aided by the concomitant development of arterial and venous systems, respond to the intake of nutrients that are required to sustain all body tissues. Essential elements — or as they are unfortunately labeled on many vitamin supplement containers 'minerals' — and all precursor molecules required for tissue production are transported by the vascular system to the osteoblasts. These cells orchestrate the form of the organ, setting out an appropriate disposition of matrix and mineral, just as the specialized cells in other tissues form the non-mineralized organs of the body. Another specialized bone cell, the osteoclast, reworks the mineralized tissue. These giant multinucleate cells resorb and reorganize the deposits during growth or repair, maintaining the architecture and the function of the organ (Jee 1983, Nijweide et al. 1986).

Bone Biochemistry

All of the separate activities that combine in the formation and reshaping of the mineralized skeletal tissues require a constant chemical flux between the tissues and the blood. Hormones and other biomolecules influence the status and the rates of exchange at all stages of tissue turnover (Canalis 1996). These biochemicals may be generated locally by the bone cells or arrive via the blood from many different organs. For example, some are derived from the food ingested, broken down, and absorbed via the intestine (Turnberg and Kumar 1993). Others, such as parathyroid hormone, derived in the thyroid gland and as-

sisted by actions of the kidney (Suki and Rouse 1996), maintain the level of ionized calcium in the blood at around 5 mg/dL and the total level of calcium in the serum between 9 and 11 mg/dL (Bowers et al. 1986). The concentration of circulating calcium in the blood and serum has as its ultimate source the composition and amount of the calcium-containing mineral per unit area of bone tissue. The hormones are the means by which the various calcium pools keep the blood and serum at the required levels. An adult human body may contain 1 kg of calcium but less than 1% is outside the skeleton (Bronnar 1997). Formation and maintenance of skeletal tissue with its important mineral constituent are homeostatic, a balance served by a series of checks and balances, a feedback system of chemical reactions that are totally integrated and internal.

The health of humans is intimately connected with the ingestion of adequate nutrients. The elements essential for all living forms — oxygen, carbon, nitrogen and hydrogen — are never in short supply (Kohn 1999), but some cations can be rate-limiting in converting these elements into the molecules of life. The list of the essential elements in addition to C, N, O, and H is presented (Table 23.1) in groups of macro, micro, and trace elements, to illustrate the present understanding of the amount of each element required daily for adequate nutrition (O'Dell and Sunde 1997).

Bone Mineral

The list contains a large proportion of the naturally occurring elements found in the Periodic Table. At the top of the list of macronutrients are Ca and P. They are essential to the formation and deposition of hydroxylapatite, $Ca_5(PO_4)_3(OH)$, the main constituent of the mineral matter of bones and teeth (Skinner 1973). Many of the elements in the list form stable solid solutions in the apatite crystal structure (Fig. 23.1), as has been shown from study of the natural occurrences of apatites in igneous, metamorphic, and sedimentary rocks (Deer et al. 1992, Skinner 1997). Carefully extracting the mineral from the organic/cellular fraction of bone shows that all of these elements occur as part of this essential inorganic constituent of the tissues. Table 23.2, an analysis of the mineral found in the dense portion (the cortex) of a leg bone of a cow, is presented to illustrate the amount of each element found in normal, heavily mineralized tissue. The

Table 23.1: Elements for adequate nutrition
Macro: Elements with daily requirements more than 100 mg   Ca, P, Na, K, Cl, Mg, S
Micro: Elements with daily requirement less than 1 mg up to 100 mg   Fe, CU, Zn, Mn, I, Mo, Se, F, Br, Cr, Co, Si
Trace: Elements whose requirements are not as yet adequately determined, but whose presence in microgram quantities is thought to be necessary. As and Pb are potentially hazardous in excess.   As, B, Sn, Ni, Ge, V, W, Pb

Table 23.2: Composition of the mineral of bovine cortical bone after ethylenediamine extraction

Element	Content (Percent by Weight)	Element	Content (Percent by Weight)
Ca	33.50 ± 0.13	P	15.53 ± 0.05
Ag	0.001	Mn	<0.001
Al	0.01–0.1	Na	1–10.0
Ba	0.001–0.01	Pb	0.001–0.01
Cr	0.001–0.01	Si	<0.001
Cu	0.01–0.1	Ni	<0.001
Fe	0.1–1.0	Sn	0.001–0.01
Mg	0.1–1.0	Sr	0.01–0.1
K	<1.0	H_2O	3.5

Spectrographic semiquantitative analysis from Spectrochemical Analysis Section, Analytical Chemical Division, National Bureau of Standards, U.S. Department of Commerce.
Other elements looked for but not detected: As, Au, B, Be, Bi, Cd, Ce, Co, Ga, Ge, Hf, Hg, In, Ir, La, Mo, Nb, Os, Pd, Pt, Rh, Ru, Sb, Sc, Ta, Te, Th, Tl, U, V, W, Y, Zn, Zr.
[Skinner (1987) p.201]

composition of human bone tissue would be very similar. Each sample of cortical bone submitted to such detailed analysis will show small differences in trace element composition based on the diet and metabolism of the specific individual.

With extreme insolubility over a range of pH, hydroxylapatite usually forms as very small crystallites, on the order of 5 x 50 micrometers, when precipitated in any biological tissues (Skinner 2000b). The crystallites, therefore, have large surface areas on which other elements and ligands can be absorbed.

The mineral in bone plays important roles in the dynamic equilibrium between the intake of elements and their availability locally and systemically. As a storehouse of elements, and an intermediary, bone mineral makes a very important chemical contribution to the viability of all tissues in the body. The deposition of hydroxylapatite in bone acts to conserve phosphorus, the most important element in the energy cycles of all living forms. We literally carry around with us our private fuel source. In light of the distinctive chemical and physical attributes of this mineral, it is not surprising that vertebrate hard tissues are mineralized with apatite (Skinner 2000a).

Strontium affords us a prime example of one of the cations that cross over from the environment into the bone mineral system. We learned about its transposition in a most direct and, indeed, urgent manner.

The radioactive isotope, ^{90}Sr, was generated during the tests on the hydrogen bomb in the early 1950s. The expectation that the radioelement would be deposited in mineralized tissues led to outcries across the U.S. and elsewhere for the military to stop testing. The focus of concern was the potential immediate, and long term, harm to children whose rapidly growing bones and teeth would incorporate the radioactive form of Sr. Numerous scientific reports were published elucidating the distribution of ^{90}Sr, as well as non-radioactive forms of Sr, in soils, milk and other foods (Hoopes 1962, Knapp 1961), as well as the uptake in humans and specifically in teeth (Bjornerstedt et al. 1957, Pennington and Jones 1961). Detailed effects of the inhalation and ingestion of ^{90}Sr-laden foods at the cellular and tissue level followed (Bustad and Goldman 1972). The data generated on the relationship of dose to response could not have been gathered without the enormous public interest in the radioactive hazard. Fortunately, the ingestion of large amounts of Ca in the children's diets prevented an excessive uptake of the radioactive form of Sr. As the tests ceased, a relieved populace moved away from worries about future cancers in their youngsters. Because of these adventitious circumstances we obtained valuable insight into the interaction of Sr and health.

Conclusions

The mineral deposited and sequestered in the skeleton reflects the composition of food ingested. Bone tissue is, in effect, a mirror of the global environment in which it forms. The importance and influence of some anions and cations on bone tissue, bone mineral, and the mechanisms of biomineralization continue to be investigated (Bronnar 1996). Studies of the many other living vertebrates, all of whom derive their nutrition directly and ultimately from the geological environment, although occasionally manipulated by humans, can also be relevant to, and increase our understanding of, the effects of a range of elements on our skeletal tissues and ultimately our total health. In addition, it seems prudent that research detailing the composition of bone mineral by biomedical scientists be correlated with the data becoming available through geochemical studies on the environment (**Plant et al.,** and **Selinus**).

References

Albright, J.A. and H.C.W. Skinner (1987) Bone: Structural organization and remodeling dynamics. Chapter 5, p.165–198 in The Scientific Basis of Orthopaedics. Edited by J.A.Albright and R. Brand. Appleton and Lange, Norwalk, CT.

Banfield, J.F and K.H. Nealson (1997) Geomicrobiology: Interactions between microbes and minerals. Reviews in Mineralogy 35, 448 pages. Mineralogical Society of America, Washington, D.C.

Bjornerstedt, R., A, Engstrom, C-J Clemendson, and A. Nelson (1957) Bone and Radiostronium. John Wiley & Sons, Inc. New York.

Bowers, G.N., C. Brassard and S. Sena (1986) Measurement of ionized calcium in serum with ion-selective electrodes. A mature technology that can meet daily service needs. Clinical Chemistry 21: 1437–1447.

Bronnar, Felix (1996) Metals in Bone: Aluminum, boron, cadmium, chromium, lead, silicon and strontium. Chapter 22 in Principles of Bone Biology Edited by Academic Press, New York.

Bronnar, Felix (1997) Calcium Chapter 2, p.13–61 in Handbook of Nutritionally Essential Mineral Elements. Edited by B.L. O'Dell and R.A. Sunde. Marcel Deker, Inc., New York

Canalis, E. (1996) Regulation of bone remodeling. p. 29–34. in Primer on Metabolic Bone Diseases and Disorders of Mineral Metabolism. 3rd Edition Edited by M. J. Favus. Lippincott-Raven Publishers, Philadelphia.

Centeno, Jose A. (this volume) Examination of tissues by chemical microscopy and biophysical techniques. P. ?.

Deer, W.A., R.A. Howie and J. Zussman (1992) An Introduction to the Rock-Forming Minerals. 2nd Edition. Apatite p. 663–669 Longman Scientific & Technical, Essex, England.

Goldman, M & L.K. Bustad (1972) (Eds) Medical Implications of Radiostrontium Exposure.. Proc. of symposium held in Davis, CA, Feb 22–24, 1971. US Atomic Energy Committee, Office of Information Services, Oak Ridge, Tenn.

Hoffman, P.E. and D.P. Schragg (2000) The snowball earth. Scientific American 282:68–75.

Hoopes, Roy (1962) A Report of Fallout in Your Food. Signet Books, New York

Hopps, H.C. and H.L. Cannon (1972) Editors of Geochemical Environment in Relation to Health and Disease. Annals New York Academy of Sciences 199, 352 pages.

Jee, W.S.S. (1983) The skeletal tissues p. 200–255 in Histology, Cell and Tissue Biology. Edited by L. Weiss, Elsevier Biomedical, New York.

Kohn, Matthew J. (1999) You are what you eat. Science, v. 283, p.335–336.

Knapp, H.A. The Effect of Deposition Rate and Cumulative Soil Level on the Concentration of Strontium −90 in U.S. Milk and Food Supplies. US Atomic Energy Commission, Division of Technical Information.

Nijweidie, P., E.H. Burger, and J.H.M. Feyen (1986) Cells of bone; proliferation, differentiation and hormonal regulation. Physiological Reviews v.66, p.855–886

O'Dell, B.L. and R.A. Sunde (1997) Editors of Handbook of Nutritionally Essential Mineral Elements. Marcel Dekker, Inc., New York

Pennington, J.A.T. & J. W. Jones (1987) Molybdenum, Nickel, Cobalt, Vanadium and Strontium in Total Diets. J. Amer. Dietetic Assn. v.897, p.1646–

Plant, Jane, David Smith, Barry Smith, and Shaun Reeder (this volume) Environmental Geochemistry for Global Sustainability.

Ross, M. & Skinner, H.C.W. (1994) Geology and human health. Geotimes v.39(1), p.13–15.

Schidlowski, M. (1988) A 3,800-million-year isotopic record of life from carbon in sedimentary rocks. Nature, v.333, p.313–318.

Selinus, Olle (this volume) Biogeochemical Monitoring in Medical Geology.

Skinner, H. Catherine W. (1972) Preparation of the mineral phase of bone using ethylene diamine extraction. Calcified Tiss. Res. v.10, p.257–268.

Skinner, H. Catherine W. (1973) Studies in the basic mineralizing system CaO-P2 O5-H2O. Calcified Tissue Research v.14, p.3–14.

Skinner, H. Catherine W. (1987) Bone: Mineral and mineralization. Chapter 6, p.198–212 in The Scientific Basis of Orthopaedics. Edited by J. A. Albright and R. Brand. Appleton and Lange, Norwalk, CT.

Skinner, H. Catherine W. (1997) The Apatite Group p. 854–863 in Dana's New Mineralogy 8th Edition. Edited by R. V. Gaines, H.C.W. Skinner, E. Foord, B. Mason and A. Rosensweig. John Wiley & Sons, New York.

Skinner, H. Catherine W. (2000a) In praise of phosphates or why vertebrates chose apatite to mineralize their skeletal elements. International Geological Review 42:232–240.

Skinner, H.Catherine W. (2000b) Minerals and human health. Chapter 11, p.383–412 in Environmental Mineralogy. Edited by D.J. Vaughan and R.A. Wogelius, European Mineralogical Union Notes in Mineralogy, Eotvos University Press, Budapest, Hungary.

Suki, W.N. and Rouse, D. (1996) Renal transport of calcium, magnesium and phosphorus. p. 472–515 in The Kidney. 5th Edition, Ed. by B.M. Brenner, WB Saunders, Philadelphia.

Turnberg, L.A. and R. Kumar (1993) Digestion and absorbtion of nutrients and vitamins. p.977–1008. in Gastrointestinal Disease: Pathophysiology, Diagnostics, Management. 5th Edition. Edited by Sleisenger, M.H., J.S. Fordtran et al., W.B. Saunders, Philadelphia.

Warren, H.V. (1954) Geology and Health. The Scientific Monthly 78(6):339–345.

24

Soil Nutrient Deficiencies
in an Area of Endemic Osteoarthritis (Mseleni Joint Disease) and Dwarfism in Maputoland, South Africa

PORTIA O. CERUTI, MARTIN FEY, AND JUSTIN POOLEY
University of Stellenbosch, Department of Soil Science, Matieland, South Africa

Figure 24.1: Woman with advanced MJD (photo courtesy Dr. W. Marasas).

Introduction

Unusually high incidences of dwarfism and an endemic osteoarthritis, called Mseleni Joint Disease (MJD), occur on the flat, sandy coastal plain of Maputaland. This rare disease begins with stiffness and pain in the joints and progresses to varying degrees of disability, with some of the afflicted requiring aid in walking and others completely immobile (Fig. 24.1). Almost 3% of local adults are dwarfs, while 38% of women and 11% of men have MJD (Fellingham 1973, Lockitch 1974). Medical studies since the 1970s have examined hematological, radiological, mycotoxicological, and genetic factors, and made comparisons with other diseases (Ballo 1996, Burger 1973, Lockitch 1973, Marasas 1986), yet have been fruitless in determining the etiology of MJD or the dwarfism. Dwarfism has been linked to Zn deficiencies in other areas and several bone-related disorders have been associated with P, Ca, and Mg deficiencies (Hidiroglou 1980). Calcium, Mg, Mn, and F deficiencies have all been speculated as possible causative factors of MJD (Fincham 1981,

1986), and the possibility of soil-derived nutrient deficiencies within this landscape is addressed.

Maputaland is located on the northeast corner of South Africa (Fig. 24.2), occupying an area about 50 by 100 km. It has a warm, subtropical climate, with summer rainfall occurring as cyclonic events, and varying from 1000 mm at the coast to 600 mm near Mseleni. Summer temperatures are high, averaging 29° C, and winters mild at 17° C. The region has high floral and faunal diversity and endemism (van Wyk 1996), and contains 15 major vegetation zones.

Geologically, Maputaland is covered with recent Quaternary sands, with several north-south paleodune cordons parallel to the coast. There is little relief and, besides the coastal dunes reaching almost 200 m above mean sea level, the average elevation is 100 m. No rivers cross the plain, but groundwater is frequently exposed at the surface, as evidenced by Lake Sibayi (Fig. 24.2) and the numerous pans in the region.

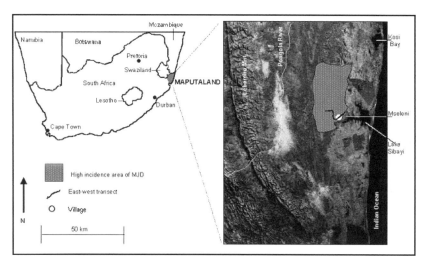

Figure 24.2: Location of Maputaland, with the transect and general high incidence area of MJD shown on a Landsat image.

Soils are mostly the Waterton family of the Fernwood form (thermic, coated Typic Quartzipsamments) (SCWG 1991, USDA 1999). These sands are inherently infertile, vary in pH from neutral to acidic, have a low cation exchange capacity, low organic matter content, and are dominated in the clay fraction by kaolinite. Fertility studies have shown soils on the western edge of the Maputaland Plain to be deficient in extractable N, P, K, Ca, Zn, and B (Lonsdale 1970).

The rural population subsists on a staple diet of locally grown maize, peanuts, sweet potatoes, and cowpeas, with no fertilizer input except cattle manure, and some food collected from the wild. The area is extremely poor, and very little food is store bought. It is speculated that nutrient deficiencies in food may contribute to malnutrition within the local community, which could then play a role in the severity of MJD, or be more directly linked to causative factors.

Experimental Procedures

Topsoil samples (0–25 cm) were taken at 1 km intervals along a 34-km roughly east-west transect through the MJD high incidence area (Pooley 1997) (Fig. 24.2). Soil chemical properties analyzed included Ambic-2 [0.25 M NH$_4$HCO$_3$ + 0.01 M (NH$_4$)$_2$-EDTA + 0.01 M NH$_4$F + Superfloc (N-127)] extractable P, K, Mn, Fe, Cu, and Zn, KCl extractable Ca and Mg, and hot water extractable B (NSWC 1990). Results of all soil analyses were interpreted in relation to

soil nutrient levels considered critical for maize (*Zea mays*), the staple diet, and below which adequate growth would not occur.

A subtractive growth trial with maize was conducted with 600 g of a composite topsoil Fernwood sand collected from a location near Mseleni, with four plants per pot (Ceruti 1999). A nutrient solution containing optimal concentrations of N, P, K, Ca, Mg, S, Mn, Fe, Cu, Zn, and B for maize was applied as a 'complete treatment' to one set of pots (in triplicate). One nutrient was then withheld in turn from further treatments, so that there were treatments representing the complete nutrient suite minus each element. Adequacy or deficiency of an element was then diagnosed from the foliar and soil nutrient concentration for the treatment in which that element was omitted, and the background level of soil fertility established. Plants were harvested at 6 weeks and oven-dried to obtain yields before an acid digestion to determine tissue nutrient concentrations (Westerman 1990). Soil nutrients determined were Bray-2 (0.03 M NH$_4$F + 0.025 M HCl) extractable P, 1 M NH$_4$-C$_2$H$_3$O$_3$ extractable K, Ca, and Mg, 0.01 M CaHPO$_4$ extractable inorganic S, 0.02 M (NH$_4$)$_2$-EDTA extractable Mn, Fe, Cu, and Zn, and CaCl$_2$ extractable B (Page et al. 1982).

Results and Discussion

The transect results show a marked variability in element concentrations across the entire length. The range of all elements varied from extremely deficient to sufficient, although Cu and Zn were extremely low across the entire transect. On average, available soil P, K, Ca, Cu, and Zn were found to be deficient, whereas Mn and B were considered borderline. Overall, Mg and Fe were not deficient. This indicates there may be pockets within the landscape of multiple deficiencies, and Cu and Zn deficiencies throughout the landscape. This is in agreement with earlier work on a nearby soil, with the addition of a severe Cu deficiency. The pattern of nutrient deficiencies did not appear to bear any sort of relationship with topographical features or vegetation patterns, and whether the deficiencies coincide with the specific locations of MJD-household gardens is unknown.

Maize yields in the subtractive trial for the minus P, K, Ca, S, and Zn treatments were all below 80% of the complete treatment, indicating multiple deficiencies. Plant tissue analysis showed deficiencies of P, K, Ca, Mg, Cu, and Zn (1.1, 3.1, 5.7, and 2.8 g kg⁻¹, and 3 and 14 mg kg⁻¹, respectively). The yield of the minus S treatment was too small for tissue analysis, and a tissue deficiency was suspected. Soil analysis confirmed deficiencies in available soil P, K, Ca, Mg, S, Cu, Zn, and B (Table 24.1). Manganese and Fe were not found to be deficient. Although on average, Mg was sufficient in the transect results, extremely deficient patches were discovered, and this was probably reflected in the composite sample.

In general, the background fertility level of this soil type is extremely low, and these severe deficiencies are most likely being transferred to the local diet. These results confirm the paucity of nutrients in local soils, specifically for two of the elements (Ca and Mg) speculated as possible causative factors of MJD and two others (P and Zn) important in bone mineralogy. Future

work is needed to confirm the subtractive trial results on a field basis, particularly as deficiency thresholds for trace elements manifest at a more advanced state of crop growth.

This work provides the first quantitative geochemical data in support of the hypothesis that a soil nutrient deficiency, or multiple deficiencies, may be responsible for MJD. Clearly the soils near Mseleni have severe nutrient deficiencies and food grown on these soils is likely to be of inferior nutritional quality. The apparent spatial heterogeneity implies that there may be a link between isolated pockets of low nutrient status and the distribution of MJD. A site-specific study is under way to spatially correlate the nutrient levels of household gardens with MJD status. Other elements such as Se, I, and F, which have been linked with health problems elsewhere, have not been considered in relation to MJD. These elements have not previously been studied in these soils but are now being investigated.

Acknowledgments:

The National Research Foundation and ESKOM's Tertiary Education Support Program are acknowledged for financial assistance.

Table 24.1: Available soil (thermic, coated Typic Quartzipsamments) nutrient concentrations from the transect study and subtractive growth trial, compared with critical levels for maize and with deficiencies indicated in bold.

Element	Range	Median	Mean ±std. dev.[b]	Critical level	Mean ±std.dev.[c]	Critical level
	mg kg⁻¹					
P	0.3 –16	1.1	**1.9** ±2.7	10	**4** ±1	7
K	3 – 114	22	**29** ±26	40	**9** ±1	40
Ca	30 – 630	210	**250** ±154	400	**347** ±45	400
Mg	10 – 60	39	49 ±32	25	**20** ±3	25
S					**11** ±2	13
Mn	0.4 – 15	4.3	5.1 ±3.6	4.7	4.6 ±0.5	3.5
Fe	4 – 36	13	17 ±7.2	5.0	43 ±2	16
Cu	0.1– 1.0	0.3	**0.4** ±0.2	1.0	**0.1** ±0.1	0.6
Zn	0.1– 1.8	0.3	**0.4** ±0.4	1.4	**0.4** ±0.1	1.0
B	0.3 – 0.9	0.4	0.5 ±0.3	0.3	**0.1**	0.5

| | Transect [a] | | | | Subtractive trial | |

(a) Results cannot be compared between transect and subtractive trials as different analytical procedures were used; the critical levels are therefore appropriate to the different extraction procedures (Ceruti 1999, Pooley 1997).
(b) Means for the transect are from 33 samples except for B, which is from 6; std. dev. = standard deviation.
(c) Means for the subtractive trial are from 3 samples except for B, which is only 1 sample; P concentration is for the minus P treatment, K concentration for minus K, etc.

References

Ballo, R. et al. 1996. Mseleni Joint Disease: A molecular genetic approach to defining the aetiology. South African Medical Journal 86:956–958.

Burger F. et al. 1973. Mseleni Joint Disease: Biochemical survey. South African Medical Journal 9:2331–2338.

Ceruti, P. 1999. Crushed rock and clay amelioration of a nutrient deficient, sandy soil of Maputaland. MSc thesis. Univ. Cape Town, RSA.

Fellingham, S., C. Elphinstone, and W. Wittmann. 1973. Mseleni Joint Disease: Background and prevalence. South African Medical Journal 47:2173–2180.

Fincham, J., S. van Rensburg, and W. Marasas. 1981. Mseleni Joint Disease: A manganese deficiency? South African Medical Journal 60:445–447.

Fincham, J., F. Hough, J. Taljaard, A. Weidemann, and C. Schutte. 1986. Mseleni Joint Disease: Part ll. Low serum calcium and magnesium levels in women. South African Medical Journal 70:740–742.

Hidiroglou, M. 1980. Zinc, copper and manganese deficiencies and the ruminant skeleton: A review. Canadian Journal of Animal Science 60:579–590.

Lockitch, G. et al. 1973. Mseleni Joint Disease: The pilot clinical survey. South African Medical Journal 47:2283–2293.

Lockitch, G. 1974. Mseleni Joint Disease: A study of the clinical and radiological aspects and possible modes of inheritance. PhD diss. Univ. Cape Town, RSA.

Lonsdale, J. 1970. Irrigation, fertility and leaching studies on a Maputa sand at Makatini. MSc thesis. Univ. Natal, RSA.

Marasas, W. and S. van Rensburg. 1986. Mycotoxicological investigations on maize and groundnuts from an endemic area of Mseleni Joint Disease in Kwazulu. South African Medical Journal 69:369–374.

NSAWC — The Non-Affiliated Soil Analysis Work Committee. 1990. Handbook of standard soil testing methods for advisory puposes. Soil Science Society of South Africa, Pretoria, RSA.

Page, A.L., R.H. Miller and D.R. Keeney (ed.). 1982. Methods of soil analysis. Part 2 _ Chemical and microbiological properties. 2nd ed. Agron. Monogr. 9. Agronomy Society of America and the Soil Science Society of America, Madison, WI.

Pooley, J. 1997. Nutrient deficiencies in soils of the Mseleni area, Kwazulu-Natal. MSc thesis. Univ. Cape Town, RSA.

SCWG - Soil Classification Working Group. 1991. Soil classification: A taxonomic system for South Africa. Department of Agricultural Development, Pretoria, RSA.

USDA, Soil Survey Staff. 1999. Soil taxonomy: A basic system of soil classification for making and interpreting soil surveys. 2nd ed. Agriculture Handbook No. 436. USDA and NRCS, Washington, DC, U.S..

van Wyk, A. 1996. Biodiversity of the Maputaland Centre. p. 198–207 *In* van der Maesan et al. (ed.) The biodiversity of African plants. Kluwer, Dordrecht, Netherlands.

Westerman, R. L. (ed.) 1990. Soil testing and plant analysis. 3rd ed. Soil Science Society of America Book Ser. No. 3. SSSA, Madison, WI.

25

Minerals in Human Blood Vessels and Their Dissolution in Vitro

MACIEJ PAWLIKOWSKI
Laboratory of Biomineralogy
Institute of Mineralogy, Petrography, and Geochemistry
University of Mining and Metallurgy, Krakow, Poland

Introduction

Cardiovascular disease knows no ethnic, national, or geographic boundaries. Men and women throughout the world can become affected by the obstruction of their arteries by cholesterol and the mineral hydroxylapatite (HA). This complex process leads to dysfunction of the arterial system and, because of the necessity of circulation of oxygenated blood, it also affects many tissues and organs. The whole process of occlusion (mineralization of the blood vessels), including precipitation and inorganic crystal formation, takes place in stages. The first stages are thought to involve cholesterol deposits (atherosclerotic plaque formation) in the interior of the vessel walls, or "intima," as it is known. The formation of hydroylapatite, or "calcification," begins with the attraction and localization of ions, mainly Ca^{2+} and PO_4^{3+}, within the arteries. The vessels become altered and lose their suppleness, effectively interfering with their function as conduits for the blood (Pawlikowski 1986, 1991a,b, 1993, Pawlikowski et al. 1994).

The initial stages of deposition can be detected with sensitive physical and chemical methods *in vivo* and with traditional laboratory methods and techniques on excised samples. In the mineralization stage, grains and crystals may become visible on heart valves as well as in the aortic tissue (Figs. 25.1–25.3). (Pawlikowski and Pfitzner 1995, 1999), and the new compounds can be identified using scanning electron microscopy and X-ray diffraction.

Reasons for the destruction of tissues and the nucleation of minerals can be attributed to allodefects and autodefects. Autodefects in vessels are those attributable to abnormalities in the component tissues in the wall or pre-existing physical conditions. For example, at arterial bifurcations, intense local trauma from the flowing blood fluid might cause changes in those regions and attract cirulating ions. Allodefects are the result of reactions between biological tissues and foreign materials, such as small particles (dust) of all sorts, including bacteria or minerals that have been inhaled and travelled from the lungs via the blood into vessels throughout the body. Alternatively, allodefects may arise from poisons produced by bacteria and viruses during infections, or by other and various chemical products, such as food preservatives, that might be part of the circulating blood. All such substances can change the electronic makeup of the artery wall and lead to lipid and apatite deposits (Pawlikowski 1991b, Pawlikowski et al. 1995, Pawlikowski and Swiderski 1995).

Christoffersen and his colleagues have investigated the dissolution of apatitic mineral matter (Christoffersen et al. 1978, Christoffersen and Christoffersen 1979, Christoffersen 1980). Other investigators have also performed similar experiments using well DASH characterized bioapatites, for example, dental enamel, that is clearly identified as HA (Gay 1962, Margolis and Moreno 1992, Voegel et al. 1983). The studies utilizing dental HA suggested that the kinetics of HA

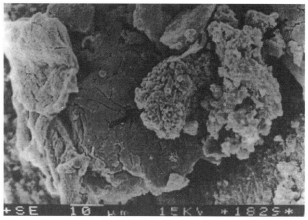

Figure 25.1: Aortic wall sampled close to the valve in Figure 2, showing accumulations of organic and inorganic deposits. SEM 1500x. Pawlikowski and Pfitzner 1999, p.62.

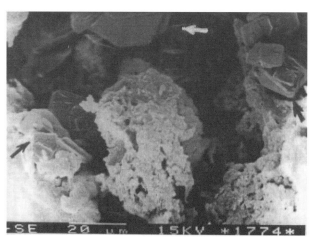

Figure 25.2: Mitral valve showing mineralization by powdery (apatitic) and crystalline (cholesterol) deposits materials. SEM 1200x. Pawlikoski and Pfitzner 1999, p. 55.

Figure 25.3: Later phase crystalline aggregates of apatitic material from a calcified aorta. SEM 5000x. Pawlikowski and Pfitzner 1999, p.24.

dissolution are controlled by surface processes and bacterial activity, so we decided to test the dissolution of the substances, especially HA and other phosphates, that crystallize in the arteries (Pawlikowski 1986, 1993; Pawlikowski et al. 1994, 1995; Pawlikowski & Pfitzner 1992, 1995; Pawlikowski and Ryskala 1991). We anticipated that different solutions would be necessary for dissolution of phosphate (decalcification), for fats or lipids, or for a mixture of inorganic-organic substances.

Experimental Investigation of the Dissolution of Hydroxyapatite in Aqueous Solutions of HCl and KOH under Static Conditions

Due to the difficulty of collecting natural HA from human vessels, HA was produced artificially following the method of Bialas et al. (1986). The synthetic HA has crystallographic and chemical features very similar to that of the natural HA in human arteries (Pawlikowski 1986, 1991b). A series of dissolution experiments on the artificially produced HA was conducted using aqueous solutions of HCl and KOH, with pH values ranging from 2.65 to 10.52. One gram of the artificial HA powder was introduced into 200 ml of the different pH solutions. The pH was measured after 30, 60, 120, 180, 210 minutes, 24 and 48 hours, and 6 days. The variations of solution pH and the length of the dissolution experiment are presented in Table 25.1.

During the experiments, it was shown that at the lower pHs, HA was not stable, that a phosphate gel was formed under the laboratory conditions at pH > 6.6, and that HA became stable between pH 7.2 and 7.6, indicating that artificial HA showed buffering capacity around the pH of normal blood. However, this capacity makes *in vivo* dissolution of the phosphatic material in the arteries difficult so a different approach, utilizing dynamic conditions, was tried.

Dissolution of HA in Water and Physiological Salts under Dynamic Conditions

In a series of experiments, synthetic HA was packed into a plastic tube connected with a system of two glass containers to a peristaltic pump. Solutions composed of distilled water and physiological salts (NaCl solution and Ringer's solution) were pumped through the system at a speed similar to that of the

Table 25.1: Results of dissolution of synthetic HA under static conditions — pH of solvents versus length of experiments

Solution Number	1	2	3	4	5	6
pH of solution – time 0	2.65	3.75	4.40	5.70	8.30	10.52
30 min	4.75	6.92	7.10	7.25	7.40	8.08
160 min	5.43	7.02	7.12	7.25	7.35	7.75
120 min	5.80	7.09	7.18	7.25	7.40	7.65
180 min	6.05	6.98	7.08	7.15	7.25	7.49
210 min	6.00	6.93	7.04	7.10	7.22	7.63
24 h	6.06	7.14	7.25	7.40	7.51	7.82
48 h	6.05	7.19	7.27	7.44	7.65	7.79
6 days	6.10	7.15	7.26	7.43	7.63	7.81

Table 25.2: Changes of content of Ca (mg/l) in solutions used for dissolving synthetic HA

Solvent	Before Dissolution	After Dissolution
Distilled water	0.151	0.443
0.9 % solution of NaCl	0.186	0.995
Ringers Solution	106.0	104.0

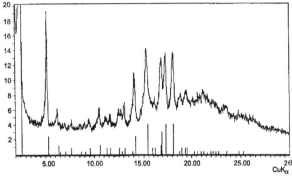

Figure 25.4: Diffraction pattern of cholesterol obtained on the crysyalline material that appears on evaporation of static ethanol dissolution of a human artery; sample from a 67-year-old man. Pawlikowski (1999) Fig.13.

blood flow in the human arteries. The HA was observed using scanning electron microscopy (SEM) before and after the experiment. The flow-through solutions were chemically analyzed by atomic absorbtion spectroscopy (AAS) and colorimetric methods before and after the traverse of the tube.

In distilled water and in the physiological solutions, the originally well-consolidated HA powder became highly porous, indicating partial dissolution. Intergranular erosion was observed, and some grains were partially dissolved. There was an increase of Ca content in the 0.9 % NaCl aqueous solution and a decrease of Ca content in Ringer's solution (Table 25.2).

A second set of experiments under similar dynamic conditions was carried out using a fragment of a mineralized human artery. The internal artery walls as well as the test solutions were investigated with the SEM and AAS before and after the experiments. Identical physiological salts to those in the previous experiment

were used. The HA-bearing artery walls, which were in a deformed state before the experiments, became less deformed afterwards.

Dissolution of Organic Substances in Arteries Using Ethanol

In another set of experiments, human coronary vessels strongly mineralized by cholesterol, were placed for 15 days in a glass vial containing pure ethanol. When the vessels were removed and the ethanol evaporated, crystals had precipitated. The crystals were examined by X-ray diffraction, and were identified as pure cholesterol, presumably part of the original arterial deposits (Fig. 25.4).

Dissolution of Organic Substances Present in Natural Arteries Using Aqueous Ethanol Solutions Under Dynamic Conditions

Fragments of another human aorta were then inserted into the pump apparatus. Ethanol diluted with distilled water (1:1) was passed over the specimen with the flow approximating the speeds of blood in natural human arteries. This experiment ran for 24 hours, after which the solvent was evaporated. The result was a deposit of yellowish and slightly transparent crystals of chloresterol up to 0.4–0.5 mm in size.

Conclusion

These experiments show that the substances that mineralize human arteries can be dissolved in a static or a flow-through system *in vitro* using acidified solutions and ethanol. The combination of lipids and mineral deposits complicates the ability to wash away these deposits. During the dissolution of hydroxyapatite using water and dissolved physiological salts under static conditions, the pH of the solutions quickly stabilized. This phenomenon took place due to the buffer features of the HA. The buffering capacity of HA was eliminated under dynamic conditions. The experiments using ethanol solutions on cholesterol and fats present in human arteries show that, as expected, the ethanol successfully dissolves them. These results suggest that future experiments using complexing solutions and a dynamic system might be successful in dissolving the mineralized portions in human arteries *in vivo*.

Acknowledgments

Investigations were financially supported by the University of Mining and Metallurgy.

References

Bialas B., Kita B., Pawlikowski M., Pospua W., 1986, Preliminary studies of synthetic hydroxyapatite and their application to bone grafts. *Mineral. Pol.*, 17, 95–107.

Christoffersen J., 1980, Kinetics of dissolution of calcium hydroxy- apatite. III. Nucleation-controlled dissolution of a poly-disperse sample of crystals. *J. Cryst. Growth.*, v.49, p. 29–44.

Christoffersen J., Christoffersen M.R.,1979, Kinetics of dissolution of calcium hydroxyapatite. II. Dissolution of non stoichiometric solutions at constant pH. *J. Cryst. Growth.*, 47, 671–679.

Christoffersen J., Christoffersen MR., Kjaergaard N., 1978, The kinetics of dissolution of calcium hydroxyapatite in water at constant pH. *J. Cryst Growth.*,43, 501–511.

Gay J.A., 1962, Kinetics of the dissolution of human dental enamel in acid. *J. Dent Res.*, 41, 633–645.

Margolis H.C., Moreno, E.C., 1992, Kinetics of hydroxyapatite dissolution in acetate, lactic, and phosphoric acid solutions. *Calcifi lIss. mt.*, 20, 137—143.

Pawlikowski, M., 1986, Mineralization of the living human organism*, Prace Mineral., 87 p.102.

Pawlikowski M., 1991a, Distribution of *it* Ca and P in human organism*, In: A. Szymariski (ed.): *Biomineralogia i biomateriaty.* PWN Warszawa, 61—66.

Pawlikowski M., 1991b, Mineralization of human arteries and lungs*, *Ibidem*, 67—83.

Pawlikowski M., 1993, *Krysztaty w organizmie cztowieka* (The crystals in the human body)*, Secesja, Krakow, 132 pp.

Pawlikowski M., 1995b, Mineralization of human tissues as an effect of pollution of environment*, *Materiaty Konferencli — Ochrona <ETH>rodowiska, 18—19 XII 1995.* AGH Krakow.

Pawlikowski, M., 1999) Preliminary results of dissolution of stances mineralizing human arteries. Arcfiwum Minerlalogicze T. Lil. Z 2, p.195—210.

Pawlikowski M., and Pfitzner R., 1992, Mineralogy of calcified heart valves*, *Folia Medica Cracoviensia*, XXXIII, 3—24.

Pawlikowski M., and Pfitzner R., 1995c, Mineralization of heart valves and coronary arteries. Materialy *2nd Baltic Sea Conference of Cardiac Interventions.* Gdafisk.

Pawlikowski, M. and Pfitzner, R., 1999, Mineralization of heart and large vessels*, Wysawnictwo Igsmie Pan., Krakow.

Pawlikowski M., Pfitzner R., Skinner C., 1995, Cholesterol-mineral concentrations of the aneurysmatic wall. *Acta Angiologica. Sapl.*, 1.

Pawlikowski M., Pfitzner R., Wachowiak J., 1994, Mineralization (calcification) of coronary arteries. *Materia Medica Polona*, 26, 3—8.

Pawlikowski M., Ryskala Z., 1991, Mineralogical-chemical characteristics of phosphate mineralization of human arteries*, *Roczniki Nauk.-Dyd. WSP w Krakowie — Prace Fizjologiczne*, 81—104.

Pawlikowski M., Swiderski R., 1995, Cholesterol-mineral concentrations in the femoral artery wall. *Acta Angiologica. SupI.*, 1, 15—16.

Voegel J.C., Gillmeth S., Frank R.M., 1983, Calcium release from powdered enamel and synthetic apatite after pretreatment with various low molecular weight organic acids. *Caries Res.*, 17, 21—220.

* References with English translation of title in Polish only.

26

Organic Compounds Derived from Pliocene Lignite and the Etiology of Balkan Endemic Nephropathy

CALIN A. TATU
Clinical Laboratory No.1, County Hospital Timisoara, Timisoara, Romania

WILLIAM H. OREM, SUSAN V.M. MAHARAJ, AND ROBERT B. FINKELMAN
U.S. Geological Survey, Reston, Virginia

DORINA DIACONITA
Geological Institute of Romania, Bucharest, Romania

GERALD L. FEDER
Florida Community College at Jacksonville, Jacksonville, Florida

DIANA N. SZILAGYI, VICTOR DUMITRASCU, VIRGIL PAUNESCU
Clinical Laboratory No.1, County Hospital Timisoara, Timisoara, Romania

Introduction

Acknowledged for the first time as a medical entity in the late 1950s, Balkan endemic nephropathy (BEN) is a chronic and irreversible kidney disease of unknown cause. BEN is geographically confined to several rural regions of central and southeastern Serbia, southwestern Romania, northwestern Bulgaria, southeastern Croatia, and parts of Bosnia and Kosovo, confined to the alluvial valleys of tributaries of the lower Danube River (Ceovic et al. 1992, Hall 1992, Tatu et al. 1998). It is estimated that several thousand people in the affected countries are currently suffering from the disease, and that thousands more will be diagnosed with BEN in the next few years.

Although no single feature is sufficient for disease diagnosis, BEN has several characteristics that allow it to be distinguished from other chronic kidney diseases: These characteristics are:

- The age of clinical onset is usually between 30 and 50 years, with a slightly higher frequency in women (female:male sex ratio is ~1.5:1), probably due to some social and genetic factors.

- There is a long subclinical "incubation" period and a rapid onset of end-stage renal disease; the only therapeutic solutions left are chronic renal dialysis or kidney transplantation.

- There is a family history of the disease, with an aggregation of the disease in certain households.

- BEN patients exhibit normal blood pressure in ~80% of the cases (a feature unusual for most other kidney diseases) and normochromic and normocytic anemia is common in BEN.

- BEN patients have lived for at least 10–20 years in one or more of the endemic villages. A slight de-

159

crease in the incidence of the disease has been noticed in the last two years in Serbia and Romania (Cukuranovic et al. 2000); however, this seems to follow the usual oscillating pattern of the disease, with lows and highs of incidence, and another epidemic outbreak is likely to occur in these areas in the future.

Many factors have been proposed as etiological agents for BEN: bacteria and viruses, heavy metals (Pb, Cd, etc.), industrial pollution, genetic factors, and even astrological influences. However, during the last decade, field and laboratory studies have progressively indicated an environmental etiology of the disease, with a major role played by the geology of the endemic settlements and their surroundings. There is a growing amount of evidence suggesting that BEN is caused by toxic organic compounds present in the well or spring drinking water from the affected villages and leached from nearby or underlying low-rank Pliocene lignite deposits (Feder et al. 1991, Orem et al. 1999). The population of villages in the endemic areas has used well/spring water almost exclusively for drinking and cooking purposes, and therefore has been directly exposed to the

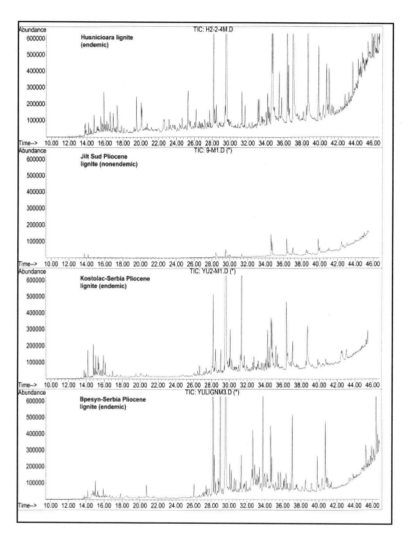

Figure 26.1: Mass spectra of three BEN endemic area Pliocene lignite samples and one nonendemic area lignite (Jilt Sud) of similar rank and age; 16 nonendemic lignites analyzed (see text) give similar spectra to Jilt Sud sample but are not shown.

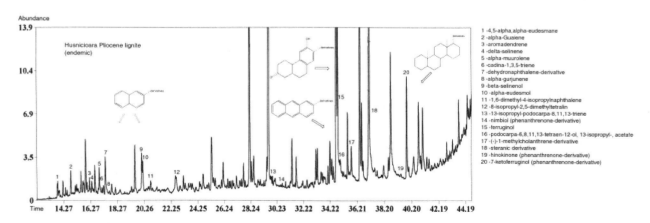

Figure 26.2: Mass spectra of an endemic Pliocene lignite with most of the compounds that could be identified labeled. The species are similar to those in the spectra illustrated in Figure 26.1.

supposed toxic compounds, with BEN developing after several decades of exposure. Another distinctive feature of BEN is its frequent association with upper urinary tract (urothelial) tumors, suggesting the action of a nephrotoxic, as well as carcinogenic, factor.

The presence of low-rank Pliocene lignite deposits is a common feature of BEN endemic sites from Romania, former Yugoslavia, and possibly Bulgaria (Tatu et al. 2000b). Our studies have been aimed at finding some unique geochemical features of these endemic area low rank coals, in order to explain the geographic restriction and the other peculiar medical features of BEN.

Materials and Methods

In order to assess the organic geochemical specificity of the Pliocene lignite from the endemic areas compared to other coals, we have analyzed by gas chromatography-mass spectrometry (GC/MS) methanol extracts of various lignite samples as follows: one Pliocene lignite (Husnicioara) from an endemic area in Romania, two Pliocene lignites (Bpesyn and Kostolac) from two different endemic areas in Serbia, 16 lignites from various BEN nonendemic coal areas of Romania, and one lignite from North Dakota. The methanol extraction was performed for 2 days at room temperature in the dark, using 1 g of coal and 2.5 ml of methanol. A volume of 1 μl of each extract was used for GC/MS analysis on an HP 6890/5973 system (Agilent Technologies, Palo Alto, CA) equipped with HP-5MS (nonpolar) column. For the mass spectra interpretation, we used the integrated Chemstation software with the NIST 1998 mass spectral library.

Since our supposition is that the water is the carrier of the BEN causing agent, we also investigated the water extractability of the coal compounds as an experimental model for their natural bioavailability. Three coal types were used for this experiment: the Pliocene lignite (Husnicioara) from the Romanian endemic area, an endemic area Pliocene lignite from Serbia (Bpesyn) and a medium-volatile bituminous coal from Maryland. A sample containing 10 g of each coal sample was subjected to soxhlet extraction with 200 ml of distilled water for 3 days. The organic compounds were extracted from the water with methylene chloride, which was subsequently concentrated by rotoevaporation and controlled evaporation in nitrogen (N_2). A volume of 1 μl of each extract was injected into a Perkin Elmer (PE Corporation, Norwalk, CT) GC 8500, with FID detector and a DB-5 column.

Results and Discussion

In earlier experiments we demonstrated that the methanol extracts of the Pliocene lignites from the endemic area have a unique geochemical signature, characterized by much more complex and abundant mass spectra, compared to higher rank coals (Tatu et al. 2000a, 2000b). ^{13}C-NMR experiments have provided similar results, showing that the endemic area Pliocene lignites are rich in aromatic and aliphatic components and high organic functionality (especially methoxy-, carbonyl-, hydroxy-, phenolic and carboxyl-groups) (Orem et al. 1999).

However, such differences seem to occur not only between the endemic Pliocene lignites and the higher rank coals but also between the former and the other lignites, collected from areas free of BEN. Mass spectra of the three endemic zone lignites and one non-endemic area lignite of similar age are shown in Figure 26.1. The complexity of the endemic Pliocene coals is obvious, and it is not matched by any of the 17 non-endemic lignites. The spectra of the endemic lignites contain aliphatic (mainly cycloalkanes/alkenes and steranic structures) and aromatic (mono- and polyaromatic terpanes, polycyclic aromatic hydrocarbons) signatures (see sample mass spectra in Fig. 26.2). Many of these have attached oxygen-based functional groups (hydroxy-, phenol-, keto-, methoxy-) and some of them contain heterocyclic nitrogen or amino groups, structural features that could make them toxic. Very few of these components are encountered in the lignite extracts from the nonendemic areas. The difference could be explained by distinctive paleogeographic coal-forming conditions during the last epoch of the Tertiary in the corresponding BEN endemic and nonendemic areas (see Tatu et al. 2000b). A lower extractability of the organic compounds from the nonendemic lignites can also explain the differences in the mass spectra. We have demonstrated in previous work (Tatu et al. 2000a) that a 3-minute methanol extraction of the endemic area Pliocene lignite is sufficient to give a high yield of organic components, the compounds being loosely bound to the coal matrix and readily extractable under the mild experimental conditions used. For higher rank coals, a much longer time is needed to extract similar amounts of compounds; this may also be true for the nonendemic area lignites analyzed.

Figure 26.3 shows that the endemic area Pliocene lignites also have a larger content of water-soluble organic compounds, compared to the higher rank coals. The individual components have not yet been identified, but it is likely that they are mainly aromatic and aliphatic and many of them may overlap with those from the methanol extracts.

Altogether, our data bring new supportive evidence for the role the Pliocene lignite geochemistry may play in the geographic restriction of the disease. However, questions related to the bioavailability of the potentially toxic coal-derived compounds are still to be answered. We are currently conducting extensive analyses of drinking water samples collected from the affected villages and laboratory experiments with various coal samples, which will shed new light on the enigma of BEN.

Acknowledgments

This work was supported through a collaborative linkage grant (CLG 975818) from NATO and through funding from the U.S. Geological Survey and the Romanian Ministry of Health.

References

Ceovic, S., A. Hrabar, M. Saric, 1992, Epidemiology of Balkan endemic nephropathy: Food and Chemical Toxicology, v. 30, no. 3, p.183–8.

Cukuranovic, R., B. Petrovic, Z. Cukuranovic, V. Stefanovic, 2000, Balkan endemic nephropathy: a decreasing incidence of the disease: Pathol Biol (Paris) 2000 Jul;48(6):558–61

Hall, P.W., 1992, Balkan endemic nephropathy: more questions than answers: Nephron, v. 62, no.1, p.1–5.

Feder, G.L., Z. Radovanovic, and R.B. Finkelman, 1991, Relationship between weathered coal deposits and the etiology of Balkan endemic nephropathy: Kidney International, v. 40, suppl. 34, p. S9–S11.

Orem, W.H., G.L. Feder, and R.B. Finkelman, 1999, A possible link between Balkan Endemic Nephropathy and the leaching of toxic organic compounds from Pliocene lignites by groundwater: International Journal of Coal Geology, v. 40, no. 2–3, p. 237–252.

Tatu, C.A., W.H. Orem, R.B. Finkelman, and G.L. Feder, 1998, The etiology of Balkan endemic nephropathy: still more questions than answers: Environmental Health Perspectives, v.106, no.11, p. 689–700.

Tatu, CA, W.H. Orem, G.L. Feder, R.B. Finkelman, D.N. Szilagyi, V. Dumitrascu, F. Margineanu, and V. Paunescu, 2000a, Additional support for the role of the Pliocene lignite derived organic compounds in the etiology of Balkan endemic nephropathy: Journal of Medicine and Biochemistry, vol. 4, no.2, p.95–101.

Tatu, C.A., W.H. Orem, G.L. Feder, V. Paunescu, V. Dumitrascu, D.N. Szilagyi, R.B. Finkelman, F. Margineanu, and F. Schneider, 2000b, Balkan endemic nephropathy etiology: a link to the geological environment: Central European Journal of Occupational and Environmental Medicine, v.6, no.2 (in press).

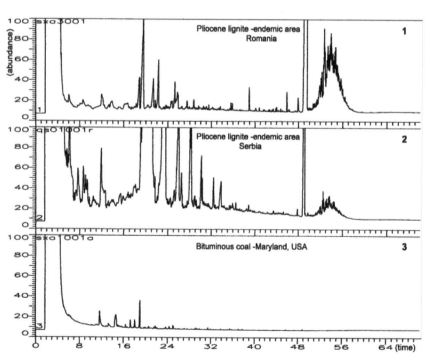

Figure 26.3: Gas chromatograms of water extracts of two BEN endemic area Pliocene lignites from Romania and Serbia, respectively, (1,2) and a bituminous coal from Maryland (3).

Summary

The contributions to this volume draw on an extensive and growing literature on chemical and physical hazards to health that are present in the natural environment. They illustrate some of the opportunities that come from combining expertise in the geologic and life sciences. Since the benefits of the natural environment in providing essential elements and nutrients are well known, the focus in this book has been on the harmful side of the relationship.

Modern science allows us to determine the basic requirements for good health and provides the tools to identify substances and places that present threats to well-being, in addition to some of the ways to ease or remove these threats. Information comes from many professions and disciplines, including medicine, dentistry, veterinary science, molecular biology, environmental toxicology, epidemiology, and medical geography. Melding the experiences of these practitioners and researchers with information from geomorphology, sedimentology, geochemistry, mineralogy, hydrology, and so on, furnishes a much broader context for solving specific issues that implicate the environment in health and disease.

The contributions assembled here demonstrate that we are now able to detect and measure physical and chemical attributes and changes in the environment over the long geological time scale, the span of human history, the periods of individual lifetimes, and, as we evaluate disease, over the weeks, days, minutes, or seconds required for the metabolism of essential or harmful materials in living forms. Medicine addresses and brings to these studies the care and repair of human health. Earth science addresses and assembles the vast range of external impacts that can affect health. By presenting an integrated view of the continuum of natural and man-made materials and the chemical and physical processes that influence human, animal, and plant health at a variety of temporal and spatial scales we may become better able to deal effectively with environmental challenges to health. A better understanding of geological processes, and their impacts on life, may also help to distinguish between anthropogenic hazards, which may be avoidable, and geogenic hazards, which may not.

The great importance of geochemical mapping, both in identifying actual dangers and as a source of insight into geographic variations in health conditions, is clear. If a particular disease is prevalent in one place compared to another, are there links to specific geochemical anomalies? Geological expertise can identify possible sources and processes that affect bioavailability, but medical insight is essential in raising the key questions, and the potential applications to local and to global human and animal health.

The importance of examining a wide range and different types of samples is also clear. Water chemistry can fluctuate with stream flow, seasonal runoff, and other short-term factors. Elemental speciation may vary in a portion of a geographic area over a restricted time interval — meaning that long-term averages, or bulk values, may be insufficient for monitoring or predicting some of the health challenges. Air quality has similar sampling problems. Monitoring by the use of sediment and soil analyses or plant rootlets, as in Sweden, deserves further consideration and extension by offering insight into the more immediate source materials and bioavailability, but requires equally difficult sampling decisions. Studies by biologists on other biota could also be added and compared to the background geochemical data already available on geological maps. And it is certainly true that not all members of an animal or human population express identical symptoms when exposed to a physical or chemical hazard within a stipulated time frame; some individuals may not be as susceptible, or as fully exposed. It is important that clinicians examining numbers of patients determine not only the critical assays for diagnosis but delineate the salient health factors that could be used to specify an environmental hazard.

Integration of the data from patients, biota, and the environment would reinforce the results of individual studies, perhaps providing additional insight into some of the 'silent' hazards, and the potential for avoiding the future promulgation of any hazard. Further, cooperative research efforts have the potential for crossovers not only of ideas and knowledge but of techniques and analytical methodologies. Many medical laboratories now identify minute amounts of specific hormones or proteins in the course of monitoring patients. Alteration of the specificity or level of production of these biomolecules on exposure to particular hazardous elements or combination of compounds is one way of detecting and determining dose response. Such investigations can provide some inferential, if not definitive, information before disease develops, and help to avoid a massive public health impact. In addition, any potentially hazardous species and their reactions can be studied utilizing laboratory animals or cultures of appropriate cell types. With such sensitive tools we should be able to extend our understandings of 'silent' geochemical hazards to the level of biochemical mechanisms.

This volume touches only a few examples of the relationships between health and the geological environment. Many areas remain for future consideration, such as bioavailability and deciphering the various pathways from sources to living creatures. We need additional integrated research, including studies within living forms, to understand the partition and sequestration of particulates and elements in certain organs and tissues that may lead to disease and death. Many of these areas have only begun to be elucidated. We offer in this book at least a few new connections, a few of the areas with potential for cooperation that can help to provide the scientific background for today's health policy decisions and actions.

Living together on this "one world," with its finite land, atmosphere, and water, we must deal with the daunting challenges of achieving sustainability on a natural background that is in some places and at some times inherently unhealthy. Deserts or frozen environments aside, certain geological situations, for example, where Hg and As occur naturally, may not be suitable for humans, animals or plants, at least not without massive intervention in the natural geochemical processes.

It is essential for the avoidance of hazards to health that we not only bear in mind the minimal daily requirements for biological existence, but also learn how to optimize the processes needed to recycle nutrients and wastes. Evaluation of diet is important for all medical diagnosis and investigation of diseases, though in many medical schools courses on nutrition are not required or may not even be available. It would be prudent to add such information in order to integrate geochemical-biochemical-medical investigations and to make reasonable connections in the continuum between human health and the lithosphere, hydrosphere, and atmosphere. Indeed, the crossover should, like other interdisciplinary fields, lead to novel hypotheses for testing.

We are convinced that geological and health science researchers will increasingly work together. The conference at which many of these chapters were first presented shows that cross-disciplinary activities are already under way involving individuals from many countries and between institutions such as geological surveys and concerned health officials in state and local governments. Some universities now offer courses with instructors from a range of scientific expertise and include individuals from schools of medicine, dentistry, epidemiology, and public health This is an obvious way to acquaint the researchers and decision-makers of the future with an integrated approach to the diversity of environmental problems and potential solutions.

How then may progress be encouraged, to facilitate cooperation and to close the gap? A key role is already being played by several organizations that are open to both geoscientists and medical practitioners. These include the International Council of Scientific Unions (ICSU), one of whose members is the International Union of Geological Sciences (IUGS) that has a Medical Geology Initiative <home.swipnet.se/medicalgeology> and the closely linked Project 454 of the International Geological Correlation Program (IGCP) now known as International Geoscience Program. There are also the Minerals in the Environment Research Network <www.mite-rn.org>, the Society for Environmental Geochemistry and Health, and the International Society for Ecosystem Health <www.ecosystemhealth.org>. Most of these organizations and programs, however, are run from the scientific rather than the medical side of the gap, and an effective single bridge to facilitate research and the exchange of information has yet to be developed. However, with help from IUGS, the U.S. Armed Forces Institute of Pathology is setting up a new registry that will link the medical/

pathology community with earth sciences, environmental, and public health professionals. Also lacking are the necessary international funding mechanisms for interdisciplinary research in geology and health, for the big intergovernmental agencies such as the UN Environment Program and the World Health Organization generally do not involve themselves in research. Perhaps the biggest challenge is to interest and draw into the dialogue more medical scientists and practitioners. We recognize the necessity to involve all sides in this quest, but our perspective sees only a part of this interdisciplinary endeavor, and therefore only a partial solution. Yet at least a start has been made.

Recognition of the importance of expanding populations, the destruction of some habitats, and the loss of biodiversity reminds us there is much work to be done to ensure a social and physical environment that can nurture people and other forms of life in our finite, albeit ever changing and dynamic, environment. We encourage all those interested in exploring the vast domain between health and earth sciences to join in the effort to identify and work toward understanding and ameliorating natural hazards to health in the geological environment.

Glossary of Medical Terms

The following sources were consulted to construct this glossary:

Dorland's Pocket Medical Dictionary, 22nd Edition (1977) W.B. Saunders Company, Philadelphia
Gray's Anatomy, 35th British Edition (1973) Rodger Warwick and Peter L. Williams (Editors) W.B. Saunders Co. Philadelphia.
Memmler, R.L., Cohen, B.J. and Wood, D.L. (1996) *The Human Body in Health and Disease*. Lippincott, Philadeplphia.
Stedman's Medical Dictionary, 2nd Edition (1990) Williams and Wilkins, Baltimore, MD.
Webster's American Dictionary, 3rd Edition (1981).

Further detail and explanations of the various medical terms or diseases can be found in *The Oxford Textbook of Medicine, the Oxford Illustrated Companion to Medicine,* and many other compendiums; for any geologic terms the *Glossary of Geology*, 4th Edition, (1997) Julia Jackson (Editor) American Geologic Institute, Alexandria, VA; and for mineral species, The 22nd Edition, (2002) of *The manual of mineral science: after James D. Dana* by Klein, C., and Hurlbut, C., John Wiley and Sons, New York, or *Dana's New Mineralogy*, 8th Edition (1997), by Gaines, R. et al., John Wiley and Sons.

ACUTE RENAL TUBULAR FAILURE an abrupt cessation of kidney function where the tubules involved in the accumulation of excess water, electrolytes, metabolic by-products, and foreign materials (possibly hazardous) that have been transported to the kidney by the blood no longer selectively secret these compounds to maintain the normal delicate and critical balance among the blood constituents.

AEROSOL a suspension of ultramicroscopic solid or liquid particles in air or gas (as smoke, fog, mist).

AMALGAM an alloy of mercury (Hg) with another metal, usually the well-known metals (excluding iron and platinum), to form a solid or liquid according to the proportion of mercury present, and often used to make tooth cement.

ARSENIASIS /ARSENISM /ARSENICOSIS /ARSENICONIOSIS/ ARSENOCONIOSIS/ ARSENOCOSIS arsenic-induced toxicology.

ASBESTOSIS a form of pneumoconiosis caused by inhalation of the fine particles of asbestos.

BIOACCUMULATION the increase in the amount of a substance, molecule, or element, brought about through the action of living creatures.

BIOREMEDIATION the use of living creatures, particularly bacteria, and their actions to extract or 'clean' a geographic area or geologic materials of actual or potential physical or chemical hazards.

BOWEN'S DISEASE a pre-cancerous lesion of the skin or mucous membranes characterized by small elevations covered by thickened horny tissue.

BRONCHITIS acute or chronic inflammation of the bronchial tubes or any part of them.

CARCINOMA a malignant tumor consisting of cells lying within the connective tissue framework of an organ, or other structure of a human or animal body.

CIRCULATORY SYSTEM the system of blood, blood vessels, lymphatics, and heart concerned with the circulation of blood and lymph.

CIRRHOSIS a chronic, progressive, disease of the liver that is the result of interstitial inflammation and characterized by an excessive formation of connective tissue followed by hardening and contraction that may result from unknown causes, toxemia, nutritional deficiency, or parasitism.

COMA a state of profound unconsciousness caused by disease, injury, or poison, from which the patient cannot be aroused, even by powerful stimuli.

CONJUNCTIVITIS inflammation of the delicate mucous membranes that line the eyelids and cover the eyeball.

CUBITAL VEIN the vein found at the inner portion of the forearm near the elbow (see *Gray's Anatomy*).

DENGUE FEVER an acute infectious disease characterized by sudden onset, headache, racking joint pain, and a rash; caused by a virus transmitted by mosquitoes of the genus Aedes chiefly in tropical and semi-tropical areas.

DEPIGMENTATION a loss of normal pigmentation.

DIABETES MELLITUS a familiar constitutional disorder of carbonate metabolism involving inadequate secretion or utilization of insulin; characterized by thirst, hunger, itching, weakness, loss of weight, and, when severe, acidosis and coma.

DIELDRIN crystalline insecticide consisting chiefly or entirely of epoxide $C_{13}H_8Cl_6O$.

DIGESTIVE SYSTEM the continuous passageway (alimentary or gastrointestinal tract) between the mouth where solid food is taken in and the anus where solid waste products are expelled from the body; includes such organs as the stomach, and small and large intestines, whose cells and their actions make it possible for nutrients to be absorbed.

DIOXIN an inflammable water soluble liquid of cyclic ether $C_4H_8O_2$ obtained from ethylene glycol and used chiefly as a solvent and dispersing agent.

DWARFISM stunted growth.

ENDOMYOCARDIAL FIBROSIS an accumulation of fibrous tissue in the muscular layer of the heart wall.

ENZYME any of a large class of complex proteinaceous substances produced by living cells and essential to life. They act like catalysts in the reversible reactions at cell temperature without themselves undergoing marked destruction in the process, but may require activators. Enzymes may be able to act outside the living organism and therefore are useful in many industrial processes.

EPIDEMIOLOGICAL STUDIES investigations that deal with the incidence, distribution, and control of disease in a population of animals, plants, or humans.

FLUOROSIS an abnormal poisoned condition caused by fluorine or its compounds.

GEOHELMINTH worm, round worm, tapeworm, or leech, that parasitizes the intestines of vertebrates.

GEOPHAGY the practice of eating earthy substances (especially clay) that is widespread and thought to represent an attempt to supply elements lacking in a scanty or unbalanced diet.

GRANULOCYTE a cell with granule-containing cytoplasm.

HEART DISEASE abnormal organic condition of the heart or of the heart and circulatory system.

HYPERKALEMIA a greater than normal concentration of potassium (K) ions in circulating blood.

HYPERTENSION abnormally high arterial blood pressure occurring without apparent or determinable prior organic changes in the tissues, possibly because of hereditary tendency, emotional tensions, fatty nutrition, or hormonal influence.

KAPOSI SYNDROME contagious disease expressed in the form of vesicles and pustules, like chicken pox, from a virus that leads to connective tissue skin cancer.

KERATIN any of a variety of sulfur-containing fibrous proteins that form the chemical basis of epidermal tissues (hair, wool, nails, horn, feathers), insoluble in most solvents and, unlike collagen and other proteins, are typically not digested by enzymes of the gastrointestinal tract.

KIDNEY one of a pair of vertebrate organs that serve, interdependent with other body systems, to regulate the volume, and balance the pH and electrolyte composition of body fluids, maintaining a balance or homeostasis for tissues. Kidneys accumulate and excrete nitrogenous waste products of metabolism such as urea, uric acid, and other substances.

KIDNEY DISEASE abnormal function of any of the multiple structures necessary for the kidney to process fluids, which may arise from obstruction, infection, toxins, or tumors, and be acute or chronic.

LEUKEMIA an acute or chronic disease of unknown origin in humans and other warm-blooded animals that involves the blood-forming organs and is characterized by abnormal increase in number of white cells (leukocytes) in tissues of the body with or without a corresponding increase in circulating blood. Classified according to the most prominent leukocyte type.

MALARIA acute or chronic disease caused by the presence of a sporozoan parasite (plasmodium) transmitted from infected to uninfected humans by the bite of an anopheline mosquito. Characterized by the destruction of blood cells from the release of toxic substances by the parasite at the end of each reproductive cycle.

MELANOSIS a condition characterized by abnormal deposits of melanin or sometimes other pigments in tissues.

MESOTHELIOMA a tumor in the mesothelial tissues mostly, although not exclusively associated with asbestos exposure and affecting the lining of the lung, the pleura, and the peritoneum.

MOLYBDENOSIS a disorder attributed to exposure to the element molybdenum (Mo).

MYCOTOXIN poison arising from the metabolism of certain fungi.

NEPHROPATHY abnormal state of the kidney, especially one associated with, or secondary to, another pathologic process.

NERVOUS SYSTEM the system in vertebrates made up of the brain, spinal cord, nerves, ganglia, and parts of receptor organs that receive and interpret stimuli and transmit impulses to effector organs.

NEURASTHENIA feeling of fatigue, worry, inadequacy, lack of zest, and interest often accompanied by headache, undersensitivity to light and noise, and disturbance of the digestion and circulation.

NEURITIS name given to an inflammatory or degenerative lesion of a nerve characterized by pain, sensory disturbance, paralysis, muscle atrophy, and impaired or lost reflexes.

NEUROTOXIN a poisonous protein complex that is present in a variety of snake venoms and exerts its principal effect as a nervous system depressant.

ORGANOCHLORINE organic compound containing chlorine.

ORGANOHALOGEN organic compound that contains fluorine, chlorine, bromine, or iodine.

OSTEOARTHITIS a degenerative disease of the joints in which the articular cartilage is eroded, either primary or secondary to trauma or other conditions, and pain and loss of function ensues. Especially common in weight-bearing joints of older people; a more highly mineralized bone may develop over time at the joint.

OSTEOSCLEROSIS abnormal mineralized tissues, usually the result of increased and haphazard deposition of mineral relative to that expected in normal bone.

PLEURAL PLAQUE deposit of dense and virtually acellular collagen (scar tissue) frequently calcified and usually described from the tissues of the chest wall (pleura) and the diaphragm (peritoneum).

PNEUMOCONIOSIS any lung disease due to permanent deposition of substantial amounts of particulate matter in the lungs.

PODOCONIOSIS disease of the feet related to the microparticles found in volcanic soils.

PULMONARY DISORDER disturbance related or associated with the lung.

RAYNAUD'S DISEASE a vascular disorder marked by recurrent spasm of the capillaries and especially those of the fingers and toes during exposure to cold. Characterized by pallor and redness in succession, commonly accompanied by pain and, in severe cases, progression to local gangrene.

RESPIRATORY DISEASE debilitation affecting the function of respiration in air-breathing vertebrates, typically a blockage that prevents transmission of oxygen to the lungs and exhalation of CO_2, but also could be malfunction of the associated nervous, circulatory, or muscle systems that aid in transport and supply of these gases.

RHAGADES fissures, cracks, or linear scars in the skin, especially around the mouth or other regions subjected to frequent movement.

SCHISTOSOMIASIS a severe disease endemic to Asia, Africa, and South America, whereby Schistosoma, a genus of parasitic worms, hosted by snails, invade through the skin of persons coming in contact with infected waters. The deposition of parasite eggs in different organs leads to symptoms and disorders peculiar to the organ infected.

SILICOSIS a condition with massive fibrosis of the lungs characterized by shortness of breath, resulting from prolonged inhalation of silica dusts, typical of some professions and construction workers.

SULFATE-REDUCING BACTERIA a unique physiological group of strictly anaerobic chemotrophic prokaryotes capable of using sulfate (SO_4) as the final electron receptor in respiration.

TACHYCARDIA relatively rapid heart action that may be physiologic (as after exercise) or pathologic.

THYROID HORMONE thyroxine or other closely related metabolically active compounds (e.g., triiodothyronine) that are stored in the thyroid gland or circulate in the blood bound to plasma proteins.

TISSUE an aggregate of cells usually of a particular kind or kinds together with their intercellular substance that forms one structural material out of which the body of a plant or animal is built up.

TRIPANOSOMIASIS tropical disease associated with arsenic exposure.

VITAMIN D any or all of several fat-soluble vitamins that are related chemically to steroids and are essential for normal tooth and bone structure. They occur in liver oils of various fishes, egg yolk, and in milk or may be synthesized by activation of sterols. Vitamin D_2 ($C_{28}H_{43}OH$) has become a dietary supplement.

Index